Learning Resource Series

AutoCAD Certification Exam Prep Workbook

Release 12

V 3.0 1993

Developed by:
Alan J. Kalameja

Produced by:
The Autodesk Learning Systems Group

Delmar Publishers Inc.™

I(T)P™

Cover design by The Autodesk Learning Systems Group

For information, address

Delmar Publishers Inc.
3 Columbia Circle, Box 15015
Albany, NY 12203-5015

Printed in the United States of America
Published simultaneously in Canada
by Nelson Canada,
a division of The Thomson Corporation

1 2 3 4 5 6 7 8 9 10   XXX   00 99 98 97 96 95 94

ISBN: 0-8273-5920-9

Contents

The AutoCAD Certification Exams

Level I
and
Level II

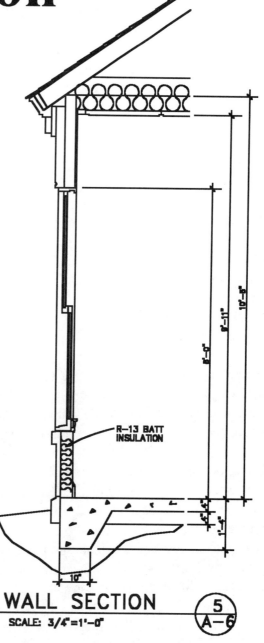

R-13 BATT
INSULATION

WALL SECTION
SCALE: 3/4"=1'-0"

5
A-6

Learning Resource Series
The Autodesk Learning Systems Group

Preface

A Letter from the AutoCAD Certification Exam Board:

Ever since the evolution of Computer-Aided Design, companies have been striving to maintain the best productive operators possible. To achieve this, these companies developed in-house tests to measure the skill and knowledge level of an operator. This test was designed around the particular application present at the company. Other individuals seeking CAD certification sought out technical colleges, community colleges, and universities for CAD instruction leading to some type of certificate. However because of differences in curriculum and grading policies, these certificates were recognized locally but not regionally and definitely not nationally. Yet another group of individuals received CAD operator training at one of the many authorized Autodesk Training Centers for AutoCAD located throughout the world. However since this type of training is performed over a three to five day period, operators were not given the time to gain a certain productivity level.

All of the above conditions lead to the development in 1990 of the AutoCAD Operator Certification Exam designed to measure the skill level of an operator and their overall knowledge of beginning and advanced AutoCAD concepts. Credit for developing this exam goes to the original exam team consisting of Robert Dugan of Moraine Valley Community College, Frank Conner of Grand Rapids Junior College, Phil Leverault of Fox Valley Technical College, Jerry Monarch and Joe Brancheau of Owens Technical College, Gary Hordemann of the Gonzaga University School of Engineering, and Ron Torrence of Everette Community College.

As the single exam format proved successful for AutoCAD Release 10 and Release 11 users, the demands of AutoCAD Release 12 resulted in a splitting of the exam. As a result, the AutoCAD Certification Exam was divided into two new exams; namely the AutoCAD Level I Certification Exam and the AutoCAD Level II Certification Exam.

Both exams last a total of 3 hours each and are broken down into two segments: a drawing segment and a general knowledge segment. The Level I drawing segment consists of 5 problems each with 5 questions. Each question in turn has 5 possible answers with only one being correct. Total time for the Level I drawing segment is 2 hours. The Level II drawing segment consists of 4 problems each with 5 questions. Each question in turn has 5 possible answers with only one being correct. Total time for the Level II drawing segment is 2 hours. Individuals must be skilled in the proper use of all AutoCAD drawing and editing commands in order to master this segment. All questions have been designed around Inquiry commands. Individuals taking these segments of the certification exams will be required to answer questions related to distances, perimeters, angles, and areas of the selected problems. Individuals attempting the Level II exam must also be skilled in the uses of the Region Modeler.

The Level I general knowledge segment consists of 75 multiple choice questions each with 5 possible answers. Total time for this segment is 1 hour. The general knowledge categories for the Level I exam include questions on blocks, drawing commands, editing commands, layers, dimensioning, beginning plotting, beginning grips, display commands, settings commands, and utility commands. The Level II general knowledge segment consists of 50 multiple choice questions each with 5 possible answers. Total time for this segment is 1 hour. The general knowledge categories for the Level II exam include questions on productivity techniques, attributes, dimension styles, advanced plotting, external references, advanced grips, model space/ paper space, and advanced utility commands.

Units 1 through 7 concentrate on preparing for the Level I Certification Exam. A series of pre and post-tests have been designed to give the individual a feel for the pace of the exam in addition to giving the user more experience in handling different test problems and questions. A unit on the use of all Inquiry commands is included followed by a series of tutorials taking the individual through a series of step-by-step tutorial problems based on Inquiry commands.

Units 9 through 13 concentrate on preparing for the Level II Certification Exam. As with the Level I exam, a series of pre and post-tests have been designed for the Level II Exam to give the individual a realistic feel for what the exam will be like. A unit on using the Region Modeler is included.

As all pre and post-tests inside of this exam prep workbook are designed to identify problem areas an operator may have, it is not intended to be used as a text book or training manual. Please refer to the AutoCAD Release 12 Reference Manual or one of the may AutoCAD text books on the market for addressing those problem areas identified by the pre and post-tests. There is also the tendency to develop certain habits in the everyday use of AutoCAD commands and options. Since the AutoCAD Certification Exam are comprehensive, the coverage of important AutoCAD features is meant to be thorough.

The members of the AutoCAD Certification Exam Board wish you success in your goals to achieve AutoCAD Level I and Level II Certification leading to a more productive individual able to compete in todays global economy.

The AutoCAD Certification Exam Board:
Bill Elliott
Alan Kalameja
John Morrison

General Information

Both AutoCAD Certification Exams are closed book; calculators, books, paper, scratch pads or other materials may not be used while the exam is in session. Scratch paper will be provided for notes or computations. No special customized menus or AutoLISP routines may be used during the exam. The Drawing Segment of both exams begin in the drawing editor with the default ACAD.DWG prototype drawing.

Both AutoCAD Certification Exams require a working knowledge of the current release of AutoCAD and test one's ability to use the software to create drawings in an efficient manner.

Since the questions designed for both Certification Exams are independent of operating systems, knowledge of DOS, UNIX, or other operating systems is not required. An individual should be familiar with the use of the current release of AutoCAD for Windows.

Questions regarding the use of AutoLISP will be few and of the most general nature. The AutoCAD Certification Exam Board has strived to develop the exam based on fundamental and general uses of 2-dimensional AutoCAD; one should not expect to be asked questions that are considered trivial or obscure in nature.

Separate certificates will be awarded for successful completion of both the Level I and Level II AutoCAD Certification Exams.

Who Should Take the Exams

This manual has been developed for the experienced user wishing to gain Level I and Level II AutoCAD Certification. An individual seeking AutoCAD Level I Certification would be classified as someone completing a beginning class on AutoCAD in addition to actually using AutoCAD in a productive environment for a period of at least 8 weeks.

An individual seeking AutoCAD Level II Certification would be classified as someone completing an advanced class on AutoCAD in addition to actually using AutoCAD in a productive environment for a period of at least 12 weeks.

Individuals with the desire to obtain either Level I and Level II Certification include, students, industrial operators, trainers, educators, and general AutoCAD users.

Exam Question Format

The Level I and Level II Certifications will be delivered electronically; this means all drawing segment and general knowledge questions will require the use of a computer for input of answers. As for the general knowledge segment, three types of questions may be asked by either the Level I or the Level II exams.

The first type of question is called "Single Answer Multiple Choice" where the individual must select the best answer out of the 5 given. For this type of question, there is only one answer that best addresses the given question. An example of the single answer multiple choice question is the following:

The command used to refresh the drawing screen without performing a regeneration is
      (A) REGEN.
      (B) REGENAUTO.
      (C) REFRESH.
      (D) REDRAW.
      (E) ZOOM-ALL.

For the question above, the correct answer is "D" since the REDRAW command is the only command of the 5 presented that performs a screen refresh without performing a regeneration.

Exam Question Format

The next type of question to be asked on either the Level I or Level II exams is a "Multiple Answer Multiple Choice" question which is similar to the single answer type except the individual must supply ALL possible answers to the question presented. An example of a multiple answer multiple choice question is the following:

Of the following list, valid Edit commands include
> **(A) TRACE.**
> **(B) STRETCH.**
> **(C) MOVE.**
> **(D) PURGE.**
> **(E) CHAMFER.**

For the question above, the correct answers are "B", "C", and "E" since STRETCH, MOVE, and CHAMFER are the only valid editing commands of the 5 presented.

The last type of question to be asked is the short answer or fill in the blank type illustrated in the following example:

The command used to set the current units of a drawing is _____.

For the question above, the correct answer is UNITS; DDUNITS would also be accepted as a correct answer.

Examples of all three question types as they will appear on actual screens have been collected and displayed in Appendix 1; The Drake Training and Testing Electronically Delivery Mechanism.

Registering for the Exams

Demand for separate Level I and Level II exams have exceeded expectations, and employers and applicants alike are seeking more guidance in measuring different levels of proficiency in AutoCAD. The complexity involved in this type of testing, coupled with the many and diverse applications for AutoCAD, pointed to the immediate need for a more sophisticated certification program. With that objective in mind, Autodesk and AutoCAD Certification Exam Board have announced that both exams will be administered by Drake Training and Technologies, a leader in testing and certification for computer professionals.

Drake has pioneered the most sophisticated administration services available using the most advanced technology. Their certification management will provide for a centralized administration and proven test development to ensure incorporation of the latest educational principles throughout the program. Both exams are offered through a computerized test delivery mechanism for immediate feedback of test results.

Testing sites are currently located at participating Authorized Autodesk Training Centers. While certification testing has already occurred in such countries as Singapore, Malaysia, and Korea, Drake's planned expansion for sites in Europe and the Far East will further establish AutoCAD Level I and Level II Certification as the worldwide standard for measuring an individuals proficiency in using AutoCAD.

To register for the exam, call Drake at:

1-800-995-EXAM

Use the above number to inquire regarding the most current fee structure for each segment of the Level I and Level II exams. When calling, you will be asked for the following information: name, social security number, mailing address, phone number, and the organization you represent.

The Future

Since both certification exams concentrate on producing two-dimensional AutoCAD drawings in a productive manner, the Level I and Level II Certification Exams in the future will constitute a Foundation Core. As a result, as an individual will receive certificates for successfully completing the Level I and Level II Certification Exams, the Foundation Core is obtained only after successfully passing both Level I and Level II Certification Exams. This procedure is illustrated in the graphic below:

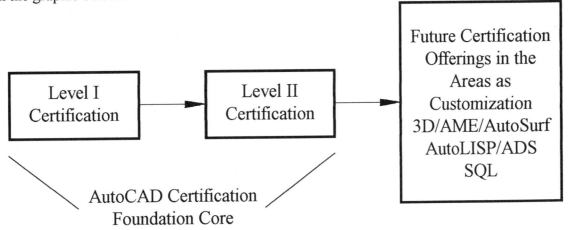

Future plans are to develop other certification exams based on such AutoCAD topics as Customization, 3D Construction, The Advanced Modeling Extension, AutoSurf/AutoMill, AutoLISP, the AutoCAD Development System, and the Structured Query Language. Once the Foundation Core is achieved, an individual may opt to enter one of these other certification areas for advanced recognition in one of these areas of specialty once they are offered in the future.

The AutoCAD Level I and Level II Certification Exams are currently being offered as the industry undergoes the development of a set of National CAD Skills. The Foundation for Industrial Modernization (FIM), a sub-group of the National Coalition for Advanced Manufacturing (NACFAM) both headquartered in Washington, DC, has been charged with developing a set of generic CAD standards for individuals with entry level skills. Future projects include the development of National CAD tests designed to assess the competency level of an individual before entering the workforce.

Acknowledgements

Creating a specialized document such as this one has been no simple task especially when jobs are potentially on the line for individuals where CAD certification is required. I would like to thank Wayne Hodgins, Grace Gallego, Barbara Bowen, Alan Jacobs, and Kirsten Rasmussen of Autodesk Inc. for their guidance in helping create this manual and Mary Beth Ray and Pamela Graul of Delmar Publishers for their technical publishing expertise. I would also like to thank the AutoCAD Certification Exam Board consisting of John Morrison of the CAD Institute in Phoenix, Arizona and Bill Elliott of Hit Return Inc., Scottsdale, Arizona for their support in putting together this document. For their part in providing feedback on the structure and objectives of both exams, I thank Mark Kurdi of Sheridan College in Ontario, Canada, Ken Flesher of the University of New Hampshire in Portsmouth, New Hampshire, J.C. Malitzke of Moraine Valley Community College in Chicago, Illinois, and Dieter Schlaepfer of Autodesk Inc. Special thanks go out to Roy Baker, Gary Crafts, Frank Dagostino, and Sid Shrum of the Trident Technical College CAD Department for their insight concerning the topic of CAD Certification and for reviewing the Pre-Test and Post-Test material. Credit for the development of the drawing problem and questions displayed on pages 126 and 127 go to the SPOCAD Centers of the Gonzaga University School of Engineering. Finally, thanks to my wife and children for their continued support and patience during this project.

About the Author

Alan J. Kalameja is the Department Head of Computer-Aided Design at Trident Technical College located in Charleston, South Carolina. He has been with the College for over 12 years and has been using AutoCAD since 1984. He directs the Authorized AutoCAD Training Center at Trident which is charged with providing industry training to local area firms. Presently, he is an Education Training Specialist with Autodesk and has authored <u>The AutoCAD Tutor for Engineering Graphics</u> published by Delmar Publishers. He joined the AutoCAD Certification Exam Board in November, 1992. He also occupies a seat on the Executive Committee charged with overseeing the development of a National CAD Standard, a project directed by the Foundation for Industrial Modernization (FIM), a sub-group of the National Coalition for Advanced Manufacturing (NACFAM) headquartered in Washington, DC.

Notes

The AutoCAD Certification Exams

Level I

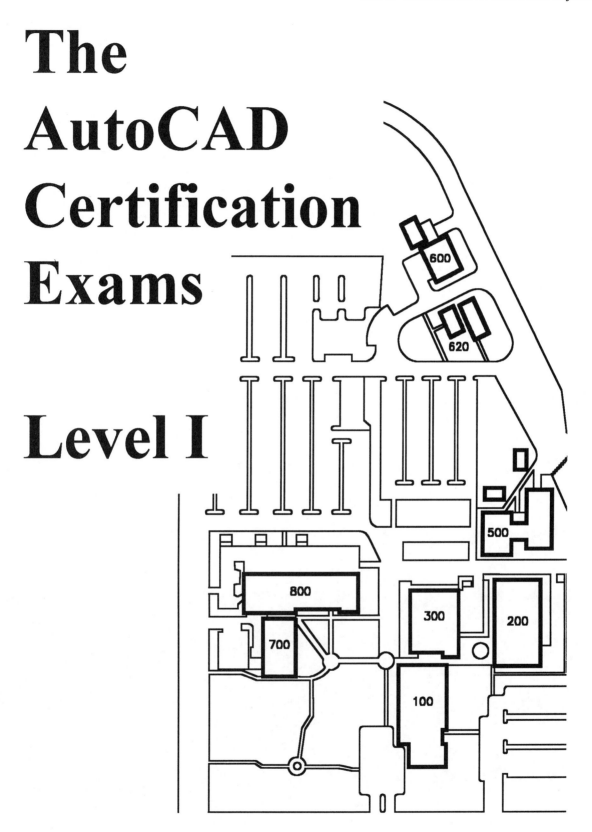

Notes

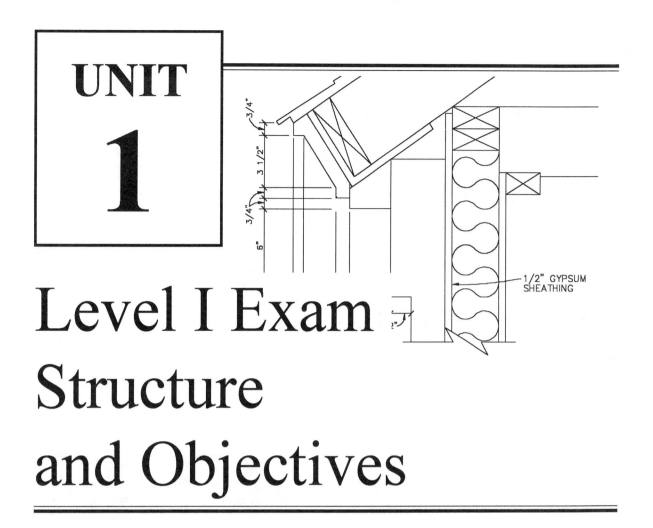

UNIT

1

Level I Exam Structure and Objectives

The AutoCAD Level I Certification Exam is divided into two parts; a drawing segment and a general knowledge segment. The drawing segment consists of 5 problems to be drawn followed by 5 questions on each problem. Inquiry commands are used to analyze each drawing; this takes the form of using such commands as Area, Dist, ID, List, and DDMODIFY. The general knowledge segment consists of 75 questions followed by 5 possible answers for each question. The individual is to select the best answer for each question.

This unit outlines the structure of the Level I Exam complete with the topics to be tested on, the number of questions per topic, and a topic percentage as it relates to the entire exam. Also, each topic is further outlined with a detailed listing of the objectives an individual must master for successfully passing the Level I Exam. Use these objectives as a study guide for determining strengths and weaknesses and what topics to concentrate on.

Level I Exam Structure Drawing Segment

The AutoCAD Level I Certification Exam is an intense 3 hour examination based on the current release of AutoCAD. Of the 3 total hours spent on the exam, 2 hours are set aside for a drawing segment and 1 hour set aside for a general AutoCAD knowledge segment. Both exam segments will be scored separately using a special statistical analysis method.

The 2 hour drawing segment consists of 5 drawing problems with 5 multiple choice questions for each problem. Future drawing segments will consist of a booklet containing 10 drawing problems where the individual will select 5 drawing problems out of the 10. Use the chart below for a breakdown on the categories of questions to be asked and the weight they carry in the Level I Exam.

Level I Exam Drawing Segment Categories	Percentage of This Segment	Number of Questions
Drawing Problem #1	20%	5
Drawing Problem #2	20%	5
Drawing Problem #3	20%	5
Drawing Problem #4	20%	5
Drawing Problem #5	20%	5
Total	100%	25

Level I Exam Structure
General Knowledge Segment

The 1 hour general AutoCAD knowledge segment of the Level I Exam consists of 75 multiple choice questions. The format of the questions will consist of single answer multiple choice questions, multiple answer multiple choice questions, and short answer (fill in the blank) questions. Use the chart below for a breakdown on the categories of questions to be asked and the weight they carry in the Level I Exam.

Level I Exam General Knowledge Segment Categories	Percentage of This Segment	Number of Questions
General AutoCAD Terminology	5%	4
Coordinate Systems	4%	3
Inquiry Commands	5%	4
Beginning Plotting	4%	3
Beginning Selection Sets	5%	4
Draw Commands	19%	14
Edit Commands	23%	16
Beginning Grips	3%	2
Beginning Dimensioning	5%	4
Block Commands	5%	4
Settings Commands	8%	6
Layers	5%	4
Display Commands	5%	4
Beginning Utility Commands	4%	3
Total	100%	75

Level I Exam Objectives

In order to successfully pass the AutoCAD Level I Certification Exam, an individual must have mastery of each of the following objectives listed below:

Objective 1.1 General AutoCAD Terminology
Know all of the following commands: New, Open, Saveas, Qsave, End, and Quit.
Know the different types of menu systems used to access AutoCAD commands.
Know the purpose of all function keys defined by AutoCAD.
Know common file types used in AutoCAD such as .DWG, .BAK, .SHX, and .DXF)

Objective 1.2 Coordinate Systems
Know how the Absolute, Relative, and Polar coordinate modes function.
Know the function of the "@" symbol.

Objective 1.3 Inquiry Commands
Know the following commands: Area, Dist, Help, ID, List, Status, and Time.

Objective 1.4 Beginning Plotting
Know the different parameters defined by the PLOT dialogue box.
Know proper plot scaling methods.

Objective 1.5 Beginning Selection Sets
Know the following modes of creating selection sets: All, Window, Crossing, Box, Fence, Last, Undo, Window, Polygon, Crossing Polygon, Previous, Add, and Remove, and Multiple.

Objective 1.6 Draw Commands
Know the following commands: Arc, Circle, Donut, Dtext, Ellipse, Line, Pline, Point, Polygon, Rectang, Sketch, and Solid.
Know the purpose of the DDPTYPE dialogue box.
Know the purpose of the Bpoly command.
Know the purpose of the BHATCH dialogue box.

Level I Exam Objectives

Objective 1.7   **Edit Commands**
Know the following commands: Array, Break, Chamfer, Change, Chprop, Copy, Divide, Erase, Explode, Extend, Fillet, Measure, Mirror, Move, Offset, Oops, Rotate, Select, Scale, Stretch, Trim, and Undo.
Know how to convert an entity into a polyline using the Pedit command.
Know how to use the Pedit-Join option to connect different segments into one polyline entity.
Know the purpose of the DDMODIFY and DDCHPROP dialogue boxes.

Objective 1.8   **Beginning Grips**
Know the purpose of the DDGRIPS dialogue box.
Know all grip modes such as Stretch, Move, Rotate, Copy, Mirror, and Copy.
Know all options associated with each grip mode.

Objective 1.9   **Beginning Dimensioning**
Know all basic dimensioning sub-commands such as Horizontal, Vertical, Aligned, Rotated, Diameter, Radius, Center, Baseline, Continuous, and Leader.
Know the ability to edit dimensions using Tedit, Newtext, and Update.
Know the effects of the Stretch, Rotate, and Scale commands on an associative dimension.
Know the use of the DDIM dialogue box in setting dimension variables.
Know the importance of manipulating the DIMSCALE variable.
Know the purpose of Ordinate dimensioning and how it is used.

Objective 1.10 **Block Commands**
Know how to create symbols using the Block and Wblock commands.
Know the Insert, Base, and Minsert commands.
Know the DDINSERT dialogue box.
Know how to redefine a block inside of a drawing.

Level I Exam Objectives Continued on the Next Page...

Level I Exam Objectives

Objective 1.11 Settings Commands
Know the following commands: Aperture, Blips, Color, Dragmod, Grid, Linetype, Limits, Ltscale, Qtext, Snap, Style, and Units.
Know all Object snap modes.
Know the purpose of the DDEMODES and DDRMODES dialogue boxes.

Objective 1.12 Layers
Know how to use all options of the DDLMODES dialogue box as they apply to to the creation of layers.
Know the affects the following commands have on layers: Color, Linetype, and Ltscale.

Objective 1.13 Display Commands
Know the following commands: Pan, Redraw, Regen, Regenauto, and View.
Know all options of the Zoom command.
Know the purpose of the DDVIEW dialogue box.

Objective 1.14 Beginning Utility Commands
Know the following commands: DXFIN, DXFOUT, Files, Purge, and Rename.
Know the purpose of the DDFILES and DDRENAME dialogue boxes.

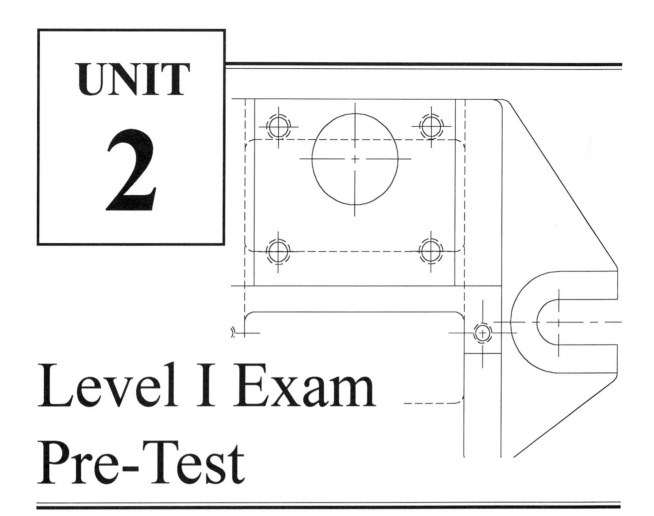

UNIT

2

Level I Exam Pre-Test

Use this unit, Level I Exam Pre-Test, to assess your current AutoCAD skill level. This Pre-Test is designed to be completed is 1 hour. It represents approximately 1/3 of the actual exam and should highlight weak areas where more work might be required to successfully pass the exam. This Level I Certification Exam Pre-Test consists of two segments:

- The first segment concentrates on drawing skills. Two problems will be drawn and the questions answered in a total of 48 minutes.

- The second segment concentrates on general AutoCAD knowledge in the form of 15 multiple choice questions. These 15 questions will be answered in 12 minutes.

Work through this Pre-Test at a good pace paying strict attention to the amount of time spent on each problem and multiple choice question. Answers for each Pre-Test question are located in Unit 7, page 138.

Notes

Level I Exam Pre-Test Section I Drawing Segment

Construct both drawing problems and answer the questions that follow each problem. The problems may be completed in any order. You should allow yourself a total of 48 minutes to complete both problems.

When both problems have been completed and time still remains, use the extra time to carefully check your answers.

Problem 1

Pattern4.Dwg

Directions for Pattern4.Dwg

Start a new drawing called Pattern4. Even though this is a metric drawing, no special limits need to be set. Keep the default setting of decimal units but change the number of decimal places past the zero from 4 to 0, (Zero). Begin constructing Pattern4 with vertex "A" at absolute coordinate (50,30). Dimensions do not have to be added to this drawing. Answer the questions on the next page regarding this drawing.

Segment Lengths
AB = 94
BC = 40
CD = 35
DE = 57
EF = 82
FG = 61
GH = 38
HJ = 85
JK = 53

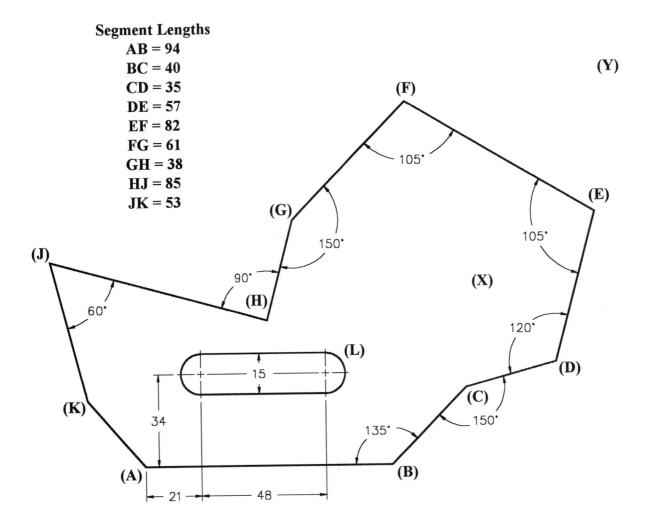

Questions for Pattern4.Dwg

Refer to the drawing of Pattern4 on the previous page to answer questions #1 through #5:

1. The total distance from the intersection at "K" to the intersection at "A" is
   - (A) 33
   - (B) 34
   - (C) 35
   - (D) 36
   - (E) 37

2. The total area of Pattern4 with the slot removed is
   - (A) 14493
   - (B) 14500
   - (C) 14529
   - (D) 14539
   - (E) 14620

3. The angle formed in the X-Y plane from the intersection at "A" to the intersection at "F" is
   - (A) 44 degrees.
   - (B) 47 degrees.
   - (C) 50 degrees.
   - (D) 53 degrees.
   - (E) 56 degrees.

4. The absolute coordinate value of the intersection at "G" is
   - (A) 104,117
   - (B) 105,118
   - (C) 106,119
   - (D) 107,120
   - (E) 108,121

5. Stretch the portion of Pattern4 around the vicinity of angle "E". Use "X" as the first corner of the crossing box. Use "Y" as the other corner. Use the endpoint of "E" as the base point of the stretching operation. For the new point, enter a polar coordinate value of 26 units in the 40 degree direction. The new degree value of angle "E" is
   - (A) 78 degrees.
   - (B) 79 degrees.
   - (C) 80 degrees.
   - (D) 81 degrees.
   - (E) 82 degrees.

Provide the answers in the spaces below:

1._____

2._____

3._____

4._____

5._____

CONTINUE ON TO THE NEXT PAGE...

Problem 2

Geneva.Dwg

Directions for Geneva.Dwg

Start a new drawing called Geneva. Keep the default setting of decimal units but change the number of decimal places past the zero from 4 to 2. Begin by constructing the 1.50 diameter hole with keyway at absolute coordinate (7.50,5.50). Dimensions do not have to be added to this drawing. Answer the questions on the next page regarding this drawing.

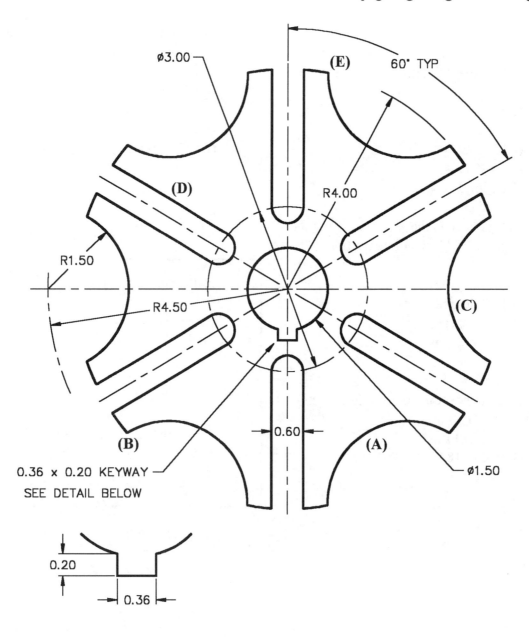

Questions for Geneva.Dwg

Refer to the drawing of Geneva on the previous page to answer questions #6 through #10:

6. The total length of arc "A" is closest to
   - (A) 3.00
   - (B) 3.10
   - (C) 3.20
   - (D) 3.30
   - (E) 3.40

7. The angle formed in the X-Y plane from the intersection at "B" to the center of arc "C" is
   - (A) 7 degrees
   - (B) 10 degrees
   - (C) 13 degrees
   - (D) 16 degrees
   - (E) 19 degrees

8. The X,Y coordinate value of the midpoint of line "D" is
   - (A) 5.27,7.13
   - (B) 5.27,7.17
   - (C) 5.23,7.13
   - (D) 5.31,7.13
   - (E) 5.31,7.09

9. The total area of the geneva with the 1.50 diameter arc and keyway removed is closest to
   - (A) 27.20
   - (B) 27.30
   - (C) 27.40
   - (D) 27.50
   - (E) 27.60

10. Use the SCALE command to reduce the geneva in size. Use 7.50,5.50 as the base point; use a scale factor of 0.83 units. The X,Y coordinate value of the intersection at "E" is
    - (A) 8.12,8.71
    - (B) 8.12,8.76
    - (C) 8.12,8.81
    - (D) 8.12,8.86
    - (E) 8.12,8.91

Provide the answers in the spaces below:

6. _____

7. _____

8. _____

9. _____

10. _____

END OF SECTION I - DO NOT PROCEED FURTHER UNTIL TOLD TO DO SO

Notes

Level I Exam Pre-Test Section II General Knowledge Segment

Answer each of the 15 multiple-choice questions on general AutoCAD knowledge. The questions may be answered in any order. It is considered good practice to answer the easier questions first. If a question seems difficult, do not waste time trying to answer it. Go on to the next question and come back to the difficult question or questions later. Be sure to provide the best possible answer for each question.

You should allow yourself a total of 12 minutes to answer all 15 multiple-choice questions.

Place the best answer in the appropriate box for each of the following questions. Unless otherwise specified, all questions are "Single Answer Multiple Choice".

11. To save a drawing file the quickest and fastest way and still remain inside of the drawing editor, use
  (A) SAVE
  (B) FSAVE
  (C) QSAVE
  (D) SAVEAS
  (E) END

12. Blocks inserted using the MINSERT command
  (A) may be exploded.
  (B) may not be exploded.
  (C) can be arranged by rows and columns similar to the ARRAY command.
  (D) can be arranged in a circular pattern similar to the ARRAY command.
  (E) only B and C.

13. To obtain the length of an arc,
  (A) use the LIST command on the arc.
  (B) use the DDMODIFY command.
  (C) convert the arc into a polyline entity and then use the LIST command.
  (D) Only A and B.
  (E) Only B and C.

14. When an entity is selected while GRIPS are enabled,
  (A) the entity highlights.
  (B) blue outlined squares are displayed at key points along the entity.
  (C) the entity highlights along with blue outlined squares being displayed at key points along the entity.
  (D) a single blue outlined square appears where the entity was originally selected at.
  (E) the entity temporarily disappears from the screen.

Question 15 is considered a "Short Answer or Fill In the Blank" question. Provide the correct answer in the space provided that satisfies the particular question.

15. The Dimension Variable that affects the overall size of all individual dimension variable settings is

_____

16. The selection set option used to select all entities in a drawing including entities on layers that have been turned off is
  (A) Entity.
  (B) All.
  (C) Add.
  (D) Screen.
  (E) Window.

17. To perform a zoom at 1/10th of the current screen size, issue the ZOOM command and type
    (A) 0.01X
    (B) 0.10X
    (C) 0.10
    (D) X0.10
    (E) 0.01

AUTOCAD

18. In the text illustration above, the special character string that causes text to be overscored is
    (A) #O
    (B) **O
    (C) ##O
    (D) %%O
    (E) %O

19. The CIRCLE-2P option prompts the user for the
    (A) endpoints of the circle's diameter.
    (B) radius of the circle.
    (C) circumference of the circle.
    (D) perimeter of the circle.
    (E) both A and B.

20. With proper linetypes loaded, the command used to modify the linetype of an entity is
    (A) DDMODIFY.
    (B) CHANGE.
    (C) DDCHPROP.
    (D) All of the above.
    (E) Only B and C.

Question 21 is considered a "Multiple-Answer Multiple-Choice question. Supply all possible answers that satisfy the particular question.

21. Valid layer names include
    (A) 1
    (B) $MECHANICAL
    (C) ISOMETRIC
    (D) FIRST   FLOOR
    (E) WIREFRAME-1

22. If a drawing is drawn in real world units, (1 to 1), and you want to plot the drawing at a scale of 1/4 its original size, the value used for the plotting scale would be
    (A) 1=1
    (B) .25=1
    (C) 1=.25
    (D) .25
    (E) 1

23. The PURGE command
    (A) must be the first command used after entering the drawing editor.
    (B) may be used at the end of a drawing session.
    (C) is used after an editing session.
    (D) may be used after performing a ZOOM command used right after entering the drawing editor.
    (E) may delete blocks currently being displayed on a drawing.

CONTINUE ON TO THE NEXT PAGE...

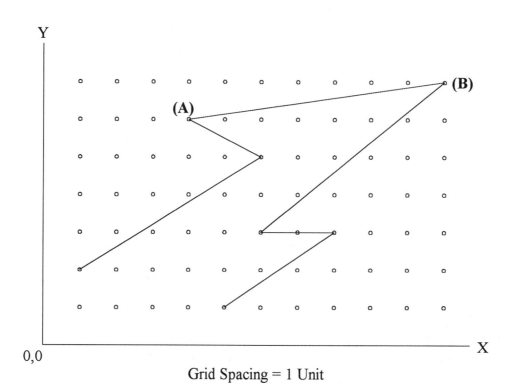

Grid Spacing = 1 Unit

24. Turning QTEXT "On" and performing a regeneration forces all text to be displayed as
    (A) rectangles.
    (B) an individual line segment for each word.
    (C) an individual line segment placed in brackets.
    (D) an individual line segment placed in parentheses.
    (E) nothing. QTEXT "On" turns all text off on the display screen.

25. In the figure above, the relative coordinates of Point "A" from Point "B" are
    (A) @-7.00,1.00
    (B) @1.00,7.00
    (C) @-7.00,-1.00
    (D) @7.00,-1.00
    (E) None of the above.

END OF SECTION II

UNIT 3

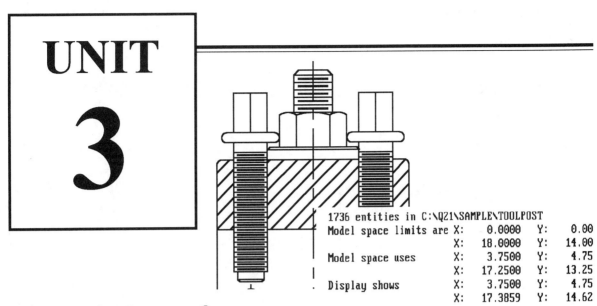

```
1736 entities in C:\Q21\SAMPLE\TOOLPOST
Model space limits are  X:   0.0000   Y:    0.00
                        X:  18.0000   Y:   14.00
Model space uses        X:   3.7500   Y:    4.75
                        X:  17.2500   Y:   13.25
Display shows           X:   3.7500   Y:    4.75
                        X:  17.3859   Y:   14.62
```

Analyzing 2-D Drawings

Completed drawings are usually plotted out and checked with scales for accuracy purposes. Depending on the thickness of pen used to perform the plot and the scale used, a range of accuracy or tolerance is assigned. A proper computer-aided design system is equipped with a series of commands to calculate distances and angles of selected entities. Surface areas may be performed on complex geometric shapes.

The next series of pages highlight all Inquiry commands and how they are used to display useful information on an entity or group of entities. The DDMODIFY command is also explained in great detail on what type of entity control it supplies to the user.

Use this information in Unit 3 to become more comfortable with all Inquiry commands and the DDMODIFY command. The user must have an excellent working knowledge of these commands in order to be successful with the first part of the AutoCAD Certification Exam, namely the Drawing Segment.

Choosing Inquiry Commands

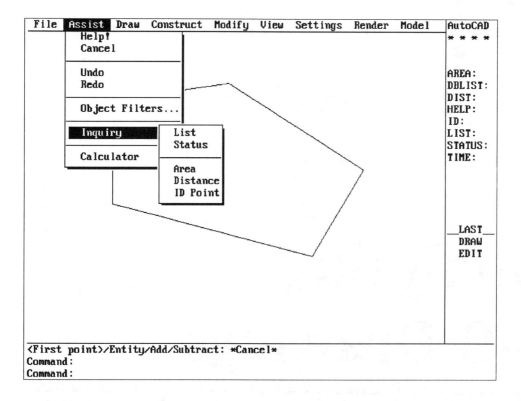

AutoCAD's Inquiry commands may be selected from the digitizing tablet, may be keyed in at the keyboard, or may be selected from the pulldown menu bar or side bar screen menus as illustrated above. The following are a listing of the Inquiry commands with a short description of each:

AREA - used to calculate the surface area given a series of points or by selecting a polyline or circle. Multiple entities may be added or subtracted to calculate the area with holes and cutouts.

DBLIST - provides a listing of all entities that make up the current drawing file.

DIST - calculates the distance between two points. Also provides the delta X,Y,Z coordi-

nate values, the angle in the X-Y plane, and the angle from the X-Y plane.

HELP - provides on-line help for any command. May be entered at the keyboard or selected from a dialog box.

ID - displays the X,Y,Z absolute coordinate of a selected point.

LIST - displays key information depending on the entity selected.

STATUS - displays important information on the current drawing.

TIME - displays the time spent in the drawing editor.

Finding the Area of an Enclosed Shape

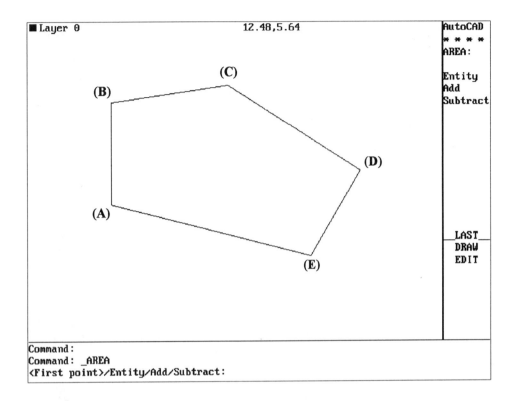

The Area command is used to calculate the area through the selection of a series of points. Select the endpoints of all vertices of the object illustrated above with the Osnap-Endpoint option. Once the first point is selected along with the remaining points in either a clockwise or counter-clockwise pattern, the command prompt "Next point:" is followed by the "Enter" key in order to calculate the area of the shape. Along with the area is a calculation of the perimeter. Use the illustrations above and to the right to gain a better understanding of the prompt sequence used for finding the area by identifying a series of points.

```
Command: AREA

<First point>/Entity/Add/Subtract: ENDP
of (Select Point "A")
Next point: ENDP
of (Select Point "B")
Next point: ENDP
of (Select Point "C")
Next point: ENDP
of (Select Point "D")
Next point: ENDP
of (Select Point "E")
Next point: ENDP
of (Select Point "A")
Next point: (Strike Enter to calculate the area)

Area = 25.00, Perimeter = 20.02
```

Finding the Area of an Enclosed Polyline or a Circle

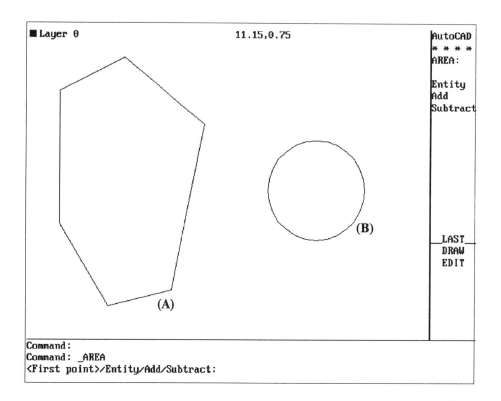

```
Command:
Command: _AREA
<First point>/Entity/Add/Subtract:
```

On the previous page, an example was given on finding the area of an enclosed shape using the Area command and identifying the corners and intersections of the enclosed area by a series of points. For a complex area, this could be a very tedious operation. As a result, the Area command has a built in Entity option which will calculate the area and perimeter on a polyline and the area and circumference on a circle. Study the illustrations above and to the right for these operations.

Finding the area of a polyline can only be accomplished if one of following are satisfied:
- The entity must have already been construced using the Pline command.
- The entity must have already been converted into a polyline using the Pedit command if originally constructed out of individual entities.

```
_AREA
<First point>/Entity/Add/Subtract: _ENTITY
Select circle or polyline: (Select "A")
Area = 24.88, Perimeter = 19.51

Command: _AREA
<First point>/Entity/Add/Subtract: _ENTITY
Select circle or polyline: (Select "B")
Area = 7.07, Circumference = 9.42
```

Finding the Area of a Surface by Subtraction

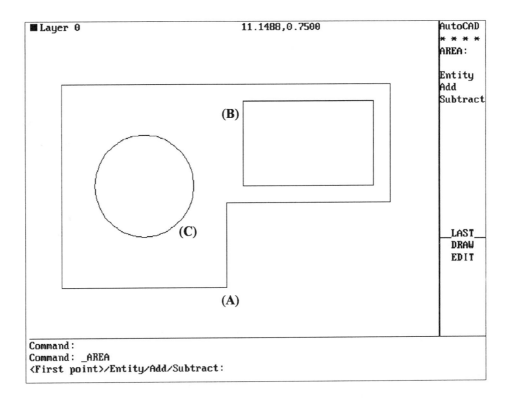

The steps used to calculate the total surface area are to first calculate the area of the outline and then subtract the entities inside of the outline. All individual entities with the exception of circles must first be converted into polylines using the Pedit command. Next, the overall area is found and added to the database using the Add mode of the Area command. "Add" mode is exited and the inner shapes are removed using the Subtract mode of the Area command. Remember, all shapes must be in the form of a circle or polyline. This means the inner shape at "B" must also be converted into a polyline using the Pedit command before calculating the area. Care must be taken when selecting the entities to subtract. If an entity is selected twice, it is subtracted twice and may yield an inaccurate area in the final calculation.

```
Command: _AREA
<First point>/Entity/Add/Subtract: _ADD
<First point>/Entity/Subtract: _ENTITY
(ADD mode) Select circle or polyline: ("A")
Area = 47.5000, Perimeter = 32.0000
Total area = 47.5000

(ADD mode) Select circle or polyline: (Enter)

<First point>/Entity/Subtract: _SUBTRACT
<First point>/Entity/Add: _ENTITY
(SUBTRACT mode) Select circle or polyline:("B")
Area = 10.0000, Perimeter = 13.0000
Total area = 37.5000

(SUBTRACT mode) Select circle or polyline:("C")
Area = 7.0686, Circumference = 9.4248
Total area = 30.4314

(SUBTRACT mode) Select circle or polyline: (Enter)

<First point>/Entity/Add: (Enter)
```

Using the Dblist (Data Base List) Command

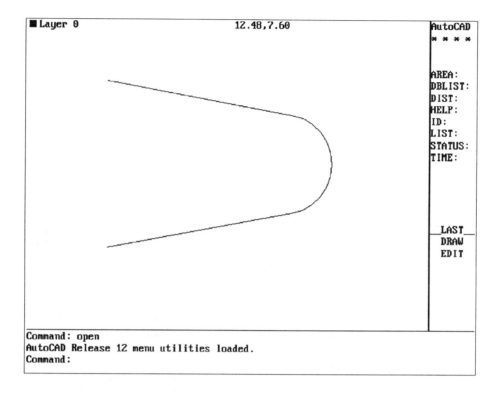

The Dblist commands simply lists all of the entities of a particular drawing that make up its database. Depending on the size of the drawing, the list of entities may be quite extensive. As a result, all entities continue to scroll over and over the text screen until the last entity lists ending the string of entities.

Depending on the number of entities in a drawing file, sometimes all of this information will not fit on the display screen. Once the screen is filled with information, strike the Enter key to continue listing more entity information.

```
       ARC        Layer: 0
                  Space: Model space
 center point, X=      8.00  Y=      5.00  Z=      0.00
 radius      1.50
   start angle    280
     end angle     80

       LINE       Layer: 0
                  Space: Model space
  from point, X=      2.50  Y=      7.50  Z=      0.00
    to point, X=      8.26  Y=      6.48  Z=      0.00
Length =     5.85,  Angle in X-Y Plane =      350
       Delta X =      5.76, Delta Y =     -1.02, Delta Z =       0.00

       LINE       Layer: 0
                  Space: Model space
  from point, X=      2.50  Y=      2.50  Z=      0.00
    to point, X=      8.26  Y=      3.52  Z=      0.00
Length =     5.85,  Angle in X-Y Plane =       10
       Delta X =      5.76, Delta Y =      1.02, Delta Z =       0.00
```

Using the Dist (Distance) Command

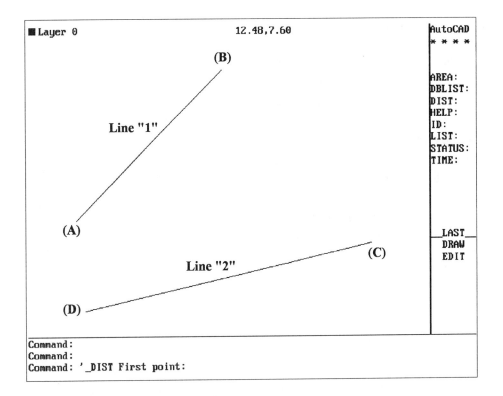

In its simplest form, the Dist command calculates the linear distance between two points on an entity whether it be the distance of a line, the distance between two points, or the distance from the quadrant of one circle to the quadrant of another circle. The following information is also supplied when using the Dist command: The angle in the X-Y plane; The angle from the X-Y plane; The delta X, Y, and Z coordinate values.

The angle in the X-Y plane is given in the current angular mode set by the Units command. The delta X, Y, and Z coordinate is a relative coordinate value taken from the first point identified by the Dist command to the second point. Using "Line 1" above as an example, "A" is identified as the first point and "B" the second. The relative distance from "A" to "B" is 5.00 units along the X axis and 4.50 units along the Y axis.

```
Command: _DIST First point: ENDP (Select at "A")
of  Second point: ENDP (Select at "B")
of
Distance = 6.73,  Angle in X-Y Plane = 42,  Angle from X-Y Plane = 0
Delta X = 5.00,  Delta Y = 4.50,   Delta Z = 0.00

Command: _DIST First point: ENDP (Select at "C")
of  Second point: ENDP (Select at "D")
of
Distance = 5.20,  Angle in X-Y Plane = 193,  Angle from X-Y Plane = 0
Delta X = -5.08,  Delta Y = -1.15,   Delta Z = 0.00
```

Interpretation of Angles using the Dist (Distance) Command

On the previous page, it was already pointed out that the Dist command yields information regarding distance, delta X,Y coordinate values, and angle information. Of particular interest is the angle in the X-Y plane formed between two points. In the illustration at the right, picking the endpoint of the line segment at "A" as the first point followed by the endpoint of the line segment at "B" as the second point displays an angle of 42 degrees. This angle is formed from an imaginary horizontal line drawn from the endpoint of the line segment at "A" in the zero direction.

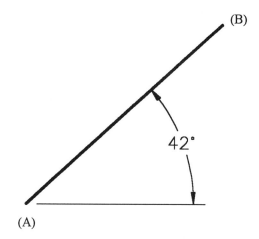

Care needs to be taken with using the Dist command to find an angle on an identical line segment illustrated to the right as with the example above. However, notice the two points for identifying the angle are selected differently. Using the Dist command, the endpoint of the line segment at "B" is selected as the first point followed by the endpoint of the segment at "A" for the second point. A new angle in the X-Y plane of 222 degrees is formed. In the illustration at the right, the angle is calculated by constructing a horizontal line from the endpoint at "B" the new first point of the Dist command. This horizonal line is also drawn in the zero direction. Notice the the relationship of the line segment to the horizontal base line. Be careful identifying the endpoints of line segments when extracting angle information.

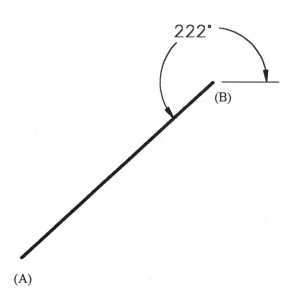

Using the ID (Identify) Command

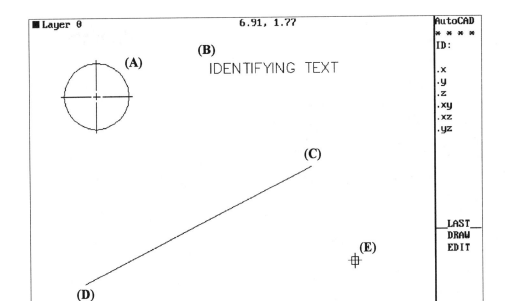

The ID command is probably one of the more staight forward of the Inquiry commands. ID stands for "Identify" and allows the user to obtain the current absolute coordinate listing of an entity.

The coordinate value of the center of the circle at "A" was found by using ID and the Osnap-Center mode; the coordinate value of the starting point of text string "B" was found using ID and the Osnap-Insert mode; the coordinate value of the endpoint of line segment "C" was found using ID and the Osnap-Endpoint mode; the coordinate value of the midpoint of the line segment at "D" was found by using ID and the Osnap-Midpoint mode; and the coordinate value of the current position of point "E" was found by using ID and the Osnap-Node mode.

```
Command: ID
Point: CEN    (Select circle "A")
of   X = 2.00      Y = 7.00      Z = 0.00

Command: ID
Point: INS    (Select text at "B")
of   X = 5.54      Y = 7.67      Z = 0.00

Command: ID
Point: END    (Select line at "C")
of   X = 8.63      Y = 4.83      Z = 0.00

Command: ID
Point: MID    (Select line at "D")
of   X = 5.13      Y = 3.08      Z = 0.00

Command: ID
Point: NOD    (Select point at "E")
of   X = 9.98      Y = 1.98      Z = 0.00
```

Using the DDMODIFY Dialogue Box

The DDMODIFY command allows the user to select an entity and display its properties in a dialogue box on the screen. Choose this command by selecting "Entity" from the "Modify" section of the pull-down menu area. Entering DDMODIFY at the command prompt is yet another way to access this command. A typical DDMODIFY dialogue box consists of the following properties: Color; Linetype; Layer; Thickness.

Below and on the next page is a brief description of the additional dialogue boxes directly related to DDMODIFY, namely Color..., Linetype..., Layer..., and Thickness.

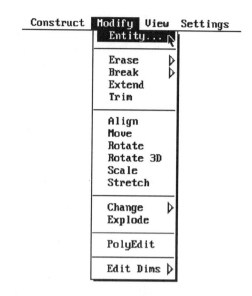

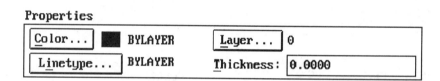

Changing the Color of an Entity

Under the "Properties" section of the DDMODIFY dialogue box is the Color option. Choosing the color property brings up an additional dialogue box displaying the current number of colors supported by the monitor configured to use AutoCAD. Picking a different color changes the selected entity to that color. This is a quick way to change color on the fly. However, changing colors this way may affect the original color of an entity set by the Layer command. It is for this reason that color be controlled by changing to a new layer.

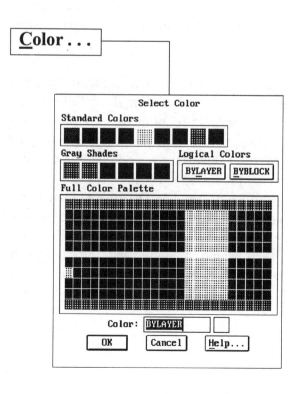

Changing the Linetype of an Entity

The "Properties" section of the DDMODIFY dialogue box also has an option for modifying the linetype of a selected entity; this new dialogue box is illustrated at the right. Choosing the linetype property brings up an additional dialogue box displaying the current linetypes loaded into the drawing. By default, only the continuous linetype displays in this dialogue box. As linetypes are loaded using the Linetype command, they will appear in this dialogue box. As with color, picking a different linetype changes the selected entity to that linetype. Also as with color, changing linetype this way may affect the original linetype of an entity set by the Layer command. It is for this reason that linetypes be controlled by changing to a new layer.

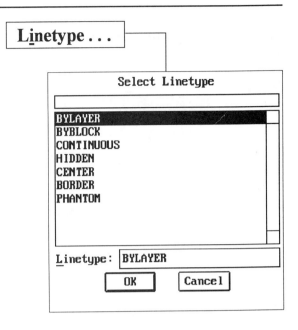

Changing the Layer of an Entity

The "Properties" section of the DDMODIFY dialogue box also has an option for modifying the layer of a selected entity illustrated below. By default, only layer 0 is created when entering a new drawing. As layers are created, they will appear in the "Select Layer" dialogue box.

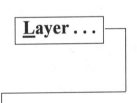

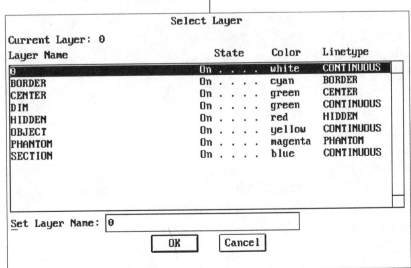

Listing the Properties of an Arc

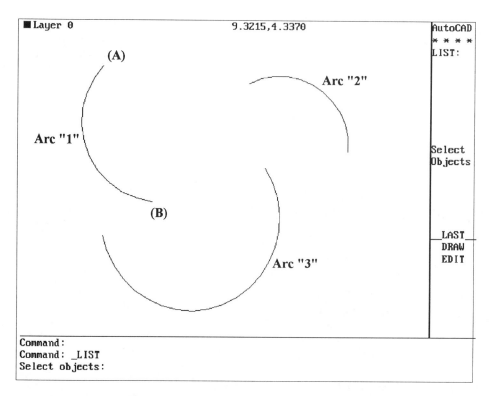

This segment of the Inquiry commands begins a study of the List command and all of the properties displayed on the particular entity listed. Using the List command on an arc provides the user with the following information: The name of the entity being listed, (Arc); The layer the entity was drawn on; Whether the entity occupies model or paper space; The center point of the arc; The radius of the arc; The starting angle of the arc in degrees; The ending angle of the arc in degrees.

The starting angle and ending angle of an arc are determined by the original construction of the arc in the counterclockwise direction. In the illustration above, listing "Arc 1" displays a starting angle at "A" of 136 degrees. Moving in a counterclockwise direction, the end angle at "B" measures 265 degrees.

```
                    Arc "1"

      ARC        Layer: 0
                 Space: Model space
center point, X=   4.2065  Y=   6.3587  Z=   0.000
radius    2.3677
 start angle    136
  end angle     265
```

```
                    Arc "2"

      ARC        Layer: 0
                 Space: Model space
center point, X=   7.9286  Y=   5.6429  Z=   0.0000
radius    2.0763
 start angle    356
  end angle     117
```

```
                    Arc "3"

      ARC        Layer: 0
                 Space: Model space
center point, X=   5.2000  Y=   3.5000  Z=   0.0000
radius    2.7459
 start angle    190
  end angle      33
```

Listing the Properties of an Arc using DDMODIFY

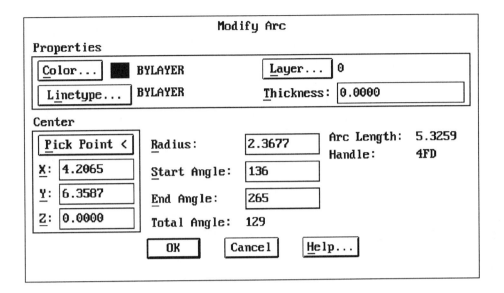

The DDMODIFY dialogue box allows for the following: Displays the center of the arc selected; Displays the radius of the arc; Displays the starting angle of the arc; Displays the ending angle of the arc; Displays the total included angle of the arc; Displays the length of the arc.

All of the above parameters of the arc that are placed inside of a box may be changed in this dialogue box and have the entity update itself after exiting the dialogue box. The illustrations at the right explain the starting angle, ending angle, total angle, and total length of the arc. The DDMODIFY command on an arc provides a more powerful listing of the entity compared with the List command on the previous page.

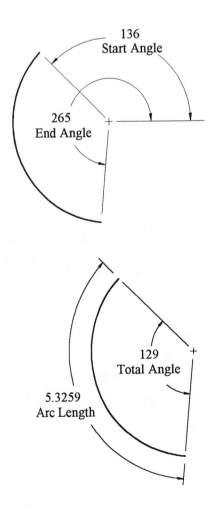

Listing the Properties of an Arc converted into a Polyline

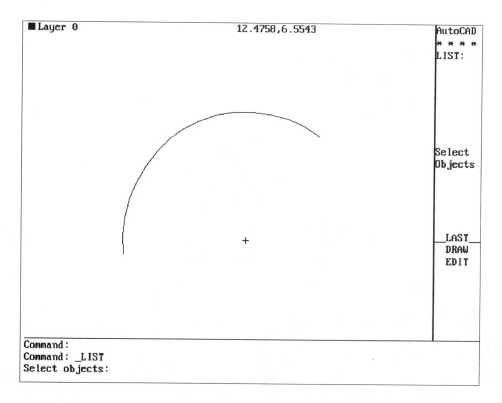

Using the List command on a polyarc entity lists the absolute coordinate values of each vertex of the polyline along with the following information: The name of the entity being listed, (Polyline); The layer the polyarc was drawn on; Whether the polyarc occupies model or paper space; The specific vertex being listed in absolute coordinates; The center point of the polyarc segment; The radius of the polyarc; The starting angle of the polyarc; The ending angle of the polyarc.

Once the last screen is displayed, the List command calculates the area occupied by a closed or open polyarc. If the polyarc is open, the total length of the polyarc segment is given.

Using the Explode command on a polyarc separates the entity into individual arc segments.

Listing the Properties of an Arc converted into a Polyline using DDMODIFY

```
                        Modify Polyline
 Properties
 [ Color... ] ▮ BYLAYER          [ Layer... ] 0

 [ Linetype... ] BYLAYER          Thickness: [ 0.0000 ]

 Polyline Type: 2D polyline      Entity Handle: None
 Vertex Listing        Fit/Smooth      Mesh              Polyline
 ┌──────────────┐      ▣ None         M:  □ Closed       □ Closed
 │ Vertex:1 [Next] │    □ Quadratic    N:  □ Closed       □ LT Gen
 │                │     □ Cubic        U: [    ]
 │ X: 9.0000      │     □ Bezier
 │ Y: 6.0000      │     □ Curve Fit    V: [    ]
 │ Z: 0.0000      │
 └──────────────┘
                 [ OK ]   [ Cancel ]   [ Help... ]
```

Using the DDMODIFY command on a polyarc lists the usual entity properties such as color, linetype, layer name, and entity thickness. This command also lists the following information: The X, Y, and Z coordinates of each polyarc vertex; The type of curve fitted to the polyarc; Whether the polyarc is closed or open; Whether to generate line type scaling per vertex.

The coordinates of all polyline vertices are identified above in the DDMODIFY dialogue box. Selecting "Next" lists the next set of coordinates. This may not be as important a function on a polyarc as on a multiple vertex polyline since the polyarc only has two vertices; one at the beginning of the polyarc and the other at the end of the polyarc. Also, the methods of fitting a curve have no affects on a polyarc.

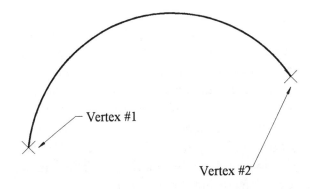

Vertex #1

Vertex #2

Listing the Properties of an Associative Dimension

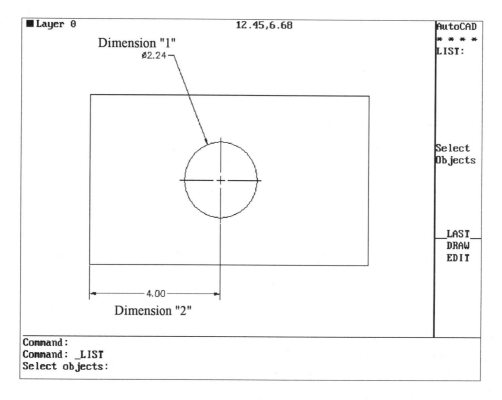

Using the List command on a diameter dimension entity lists the following information: The specific entity being listed (Dimension); The layer the dimension was drawn on; Whether the dimension occupies model or paper space; The first defining point of the leader; The second defining point of the leader; The text position (in absolute coordinates); The default text; The current dimension style.

```
                    Dimension "1"

              DIMENSION  Layer: 0
                         Space: Model space
type: diameter
        defining point: X=      6.41  Y=      3.57  Z=      0.00
defining point:         X=      5.59  Y=      5.65  Z=      0.00
leader length     2.83
default text position: X=      3.91  Y=      8.27  Z=      0.00
default text
dimension style: STANDARD
```

Using the List command on a linear dimension entity lists the following information: The specific entity being listed (Dimension); The layer the dimension was drawn on; Whether the dimension occupies model or paper space; The first extension line defining point; The second extension defining point; The text position (in absolute coordinates); The default text; The current dimension style.

```
                    Dimension "2"

              DIMENSION  Layer: 0
                         Space: Model space
type: horizontal
1st extension  defining point: X=      2.00  Y=      2.11  Z=      0.00
2nd extension  defining point: X=      6.00  Y=      3.37  Z=      0.00
dimension line defining point: X=      6.00  Y=      1.25  Z=      0.00
default text position: X=      4.00  Y=      1.25  Z=      0.00
default text
dimension style: STANDARD
```

Listing the Properties of an Associative Dimension using DDMODIFY

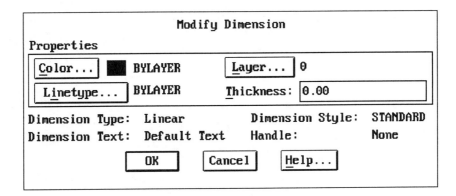

Using the DDMODIFY command on a linear dimension lists the following information: The dimension type; The dimension text; The current dimension style.

Compared with the effects DDMODIFY has on other entities, this dialogue box is similar to using the List command on a dimension.

No special tools are available to change or modify an associative except for the following standard items displayed at the top of the dialogue box: Modifying the Color of the associative dimension; Modifying the Linetype of the associative dimension; Modifying the Layer of the associative dimension; Modifying the Thickness of an associative dimension.

Listing the Properties of a Block

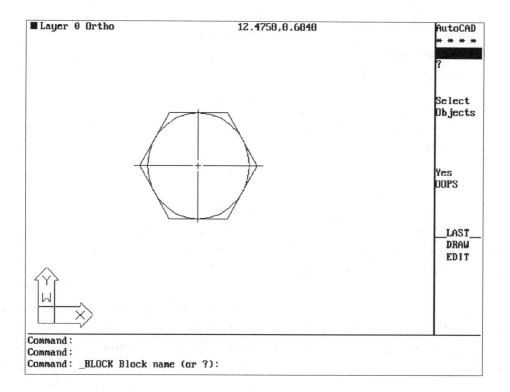

Using the List command on an block provides the user with the following information: The name of the entity being listed, (Block Reference); The layer the block was inserted on; Whether the block occupies model or paper space; The name of the block which in this case is "Bolthead"; The insertion point of the block in X, Y, and Z coordinates; The X, Y, and Z scale factors of the block; The rotation angle of the block.

```
BLOCK REFERENCE  Layer: 0
             Space: Model space
     BOLTHEAD
    at point, X=  5.7756  Y=  4.7283  Z=  0.0000
     X scale factor   1.0000
     Y scale factor   1.0000
rotation angle     0
     Z scale factor   1.0000
```

Illustrated at the right is an example of the "Bolthead" block. Point "A" identifies the insertion point of the block or where the block is inserted in relation to. All scaling and rotation values are referenced by the insertion point. The insertion point of a block is also used for reference where mutltiple block insertions a specific distances are required. This will be explained further on page 52.

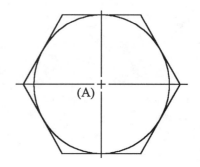

Listing the Properties of a Block using DDMODIFY

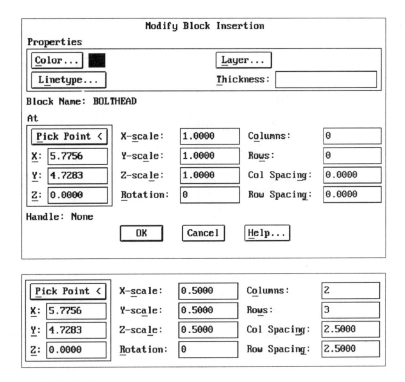

Control of blocks are enhanced through the use of the DDMODIFY command which displays the dialogue box illustrated above. Using this dialogue box allows the user to dynamically change each of the following items: The insertion point of the block; The X, Y, and Z scale values of the block; The rotation angle of the block.

Using DDMODIFY to control multiple block insertions is illustrated in the second illustration above. With the insertion point of the block at the center of the bolt head, 2 columns and 3 rows are specified. Both spacings inbetween columns and rows are 2.50 units. This creates the arrangement of bolt heads at the right.

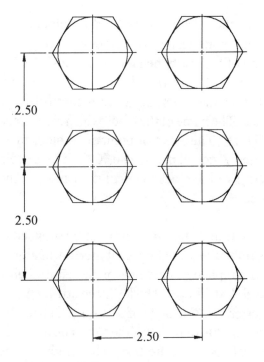

Listing the Properties of a Block with Attributes

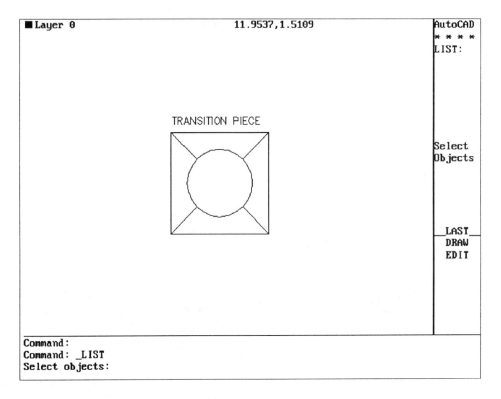

Using the List command on an block with attributes information provides following: The name of the entity being listed, (Block Reference); The layer the block was inserted on; Whether the block occupies model or paper space; The name of the block which in this case is "Duct"; The insertion point of the block in X, Y, and Z coordinates; The X, Y, and Z scale factors of the block; The rotation angle of the block.

In addition to the block information supplied at the right, the following attribute information is also supplied: The name of the entity being listed, (Attribute); The layer the attribute was inserted on; The style of text for the attribute; The font file of the attribute text; The justification of the attribute text; The height of the attribute text; The attribute value; The attribute tag; The rotation angle, width factor, obliquing angle of the attribute text.

```
BLOCK REFERENCE  Layer: 0
          Space: Model space

     DUCT
     at point, X=  6.0000 Y=  4.5000 Z=  0.0000
     X scale factor    1.0000
     Y scale factor    1.0000
rotation angle      0
     Z scale factor    1.0000

     ATTRIBUTE  Layer: 0
             Space: Model space
   Style = STANDARD    Font file = SIMPLEX
   center point, X=  5.9976 Y=  6.2349 Z=  0.0000
  height    0.2000
    value TRANSITION PIECE
      tag PART_NAME
 rotation angle      0
     width scale factor    1.0000
 obliquing angle      0
    flags normal
generation normal

       END SEQUENCE  Layer: 0
             Space: Model space
```

Listing the Properties of a Block with Attributes using DDMODIFY

```
                    Modify Block Insertion
  Properties
  |Color...| ■                    |Layer...|
  |Linetype...|                   Thickness: [        ]

  Block Name: DUCT
  At
  |Pick Point <|  X-scale: [1.0000]   Columns:  [0      ]
  X: [6.0000]     Y-scale: [1.0000]   Rows:     [0      ]
  Y: [4.5000]     Z-scale: [1.0000]   Col Spacing: [0.0000]
  Z: [0.0000]     Rotation: [0    ]   Row Spacing: [0.0000]
  Handle: None
                  [  OK  ]  [ Cancel ]  [ Help... ]
```

Control of blocks with attributes are enhanced similar to blocks without attributes through the use of the DDMODIFY command which displays the dialogue box illustrated above. Using this dialogue box allows the user to dynamically change each of the following items: The insertion point of the block; The X, Y, and Z scale values of the block; The rotation angle of the block.

This dialogue box also controls multiple inserts of a block by allowing the user to change each of the following items: The number of columns; The number of rows; A value for the spacing inbetween columns; A value for the spacing inbetween rows.

With the insertion point of the block at the center of the duct, 2 columns and 2 rows are specified. Spacing between rows is 5.00 units while spacing between columns is 4.00 units. The results are illustrated at the right. Notice since the attribute information is part of the block, it is inserted along with the block at multiple intervals.

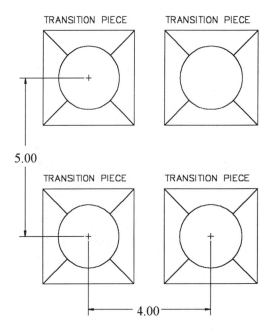

Listing the Properties of a Circle

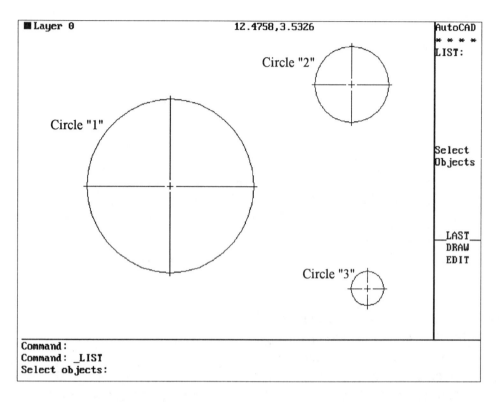

Using the List command on a circle displays the following information: The name of the entity being listed, (Circle); The layer the circle was drawn on; Whether the circle occupies model or paper space; The center point of the circle; The radius of the circle; The circumference of the circle; The area of the circle.

As with all circle properties such as center point, radius, circumference, and area, the number of decimal places of accuracy is set by the Units command.

Circle "1"

```
CIRCLE     Layer: 0
           Space: Model space
      center point, X=   4.5000  Y=   4.5000  Z=   0.0000
         radius    2.5495
    circumference  16.0190
         area     20.4204
```

Circle "2"

```
CIRCLE     Layer: 0
           Space: Model space
      center point, X=  10.0000  Y=   7.5000  Z=   0.0000
         radius    1.1180
    circumference   7.0248
         area      3.9270
```

Circle "3"

```
CIRCLE     Layer: 0
           Space: Model space
      center point, X=  10.5000  Y=   1.5000  Z=   0.0000
         radius    0.5000
    circumference   3.1416
         area      0.7854
```

Listing the Properties of a Circle using DDMODIFY

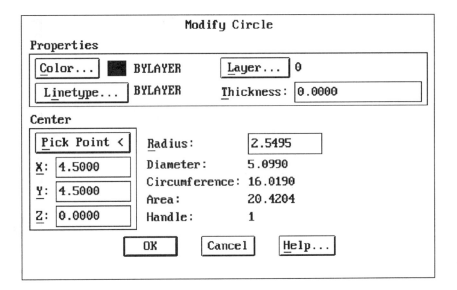

Using the DDMODIFY command on a circle lists the identical properties to change as with such entities as lines and arcs. Color, linetype, and layer name all bring up additional dialogue boxes to make it easier to edit the entity listed.

In addition to the above properties, the following additional parameters are listed of the circle: Center point of the circle; Radius of the circle; Diameter of the circle; Circumference of the circle; Area of the circle.

A new center point may be selected by selecting the box "Pick point" or by entering a new coordinate value in the appropriate X, Y, and/or Z boxes as in the first illustration at the right. Also the radius of the circle may be changed by editing the value in the radius box.

The diameter, circumference, and area values are all listed but may not be changed from this dialogue box. Notice also that when changing the circle radius, the diameter, circumference, and area values update themselves based on the new radius as in the second illustration at the right.

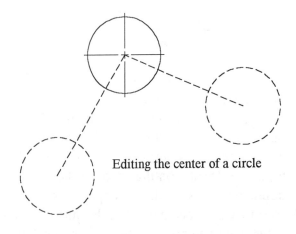

Editing the center of a circle

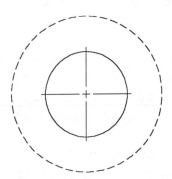

Editing the radius of a circle

Listing the Properties of a Donut

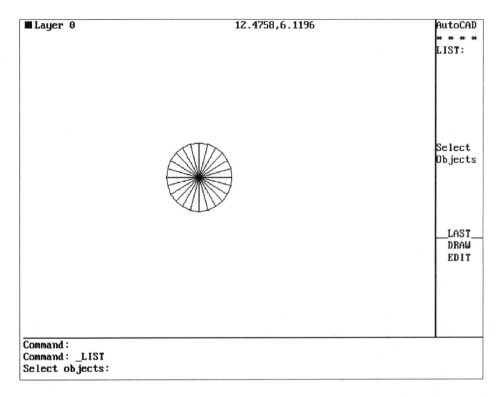

```
■Layer 0                    12.4758,6.1196          AutoCAD
                                                    * * * *
                                                    LIST:


                                                    Select
                                                    Objects


                                                    LAST
                                                    DRAW
                                                    EDIT


Command:
Command: _LIST
Select objects:
```

Using the List command on a donut entity lists the absolute coordinate values of each vertex since the donut is actually constructed as a polyline with the following information: The name of the entity being listed, (Polyline); The layer the donut was drawn on; Whether the donut occupies model or paper space; The specific vertex being listed in absolute coordinates; The center point of the polyarc segment; The radius of the polyarc; The starting angle of the polyarc; The ending angle of the polyarc.

Since the donut is a closed polyline entity, the List command calculates the area and perimeter. If the donut is open, the total area and length of the donut segment is given.

To reduce the amount of Regen time required to display donuts, use the Fill command and turn off the filled in part of a donut.

```
ending width    1.0000
       bulge    1.0000
      center X=  5.3458  Y=   4.7169  Z=   0.0000
      radius    0.5000
 start angle    180
   end angle    0

            VERTEX    Layer: 0
                      Space: Model space
          at point, X=  5.8458  Y=  4.7169  Z=   0.0000
starting width    1.0000
 ending width     1.0000
       bulge      1.0000
      center X=   5.3458  Y=   4.7169  Z=   0.0000
      radius     0.5000
 start angle     0
   end angle     180

            END SEQUENCE  Layer: 0
                          Space: Model space
        area       0.7854
   perimeter       3.1416
```

Open Donut

Listing the Properties of a Donut using DDMODIFY

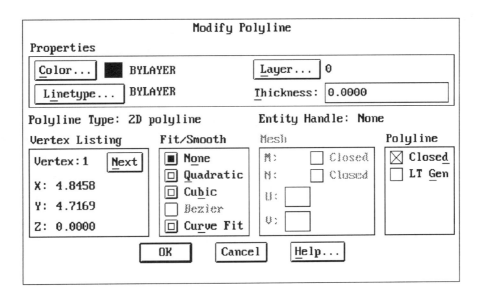

Using the DDMODIFY command on a donut lists the usual entity properties such as color, linetype, layer name, and entity thickness. This command also lists the following information: The X, Y, and Z coordinates of each donut vertex; The type of curve fitted to the donut; Whether the donut is closed or open; Whether to generate line type scaling per vertex

By default, a donut has normal curve generation which displays the donut as a filled in circular polyline. Selecting Quadratic, Cubic, and Fit Curve displays the results similar to the upper illustration at the right. As a result, any special curve generations of a donut should not be used.

Using the Explode command on a closed donut separates the entity into two individual arc segments in the second illustration at the right. The center dividing line is not present when the donut is exploded and is used only for illustrative purposes. Using Explode on an open donut converts the entity into a single arc segment.

Quadratic, Cubic, and Curve Fit on a Donut

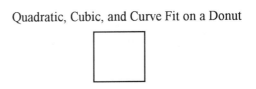

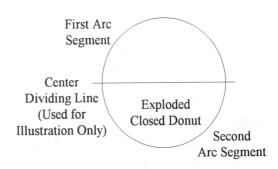

First Arc Segment

Center Dividing Line (Used for Illustration Only)

Exploded Closed Donut

Second Arc Segment

Exploded Open Donut

Listing the Properties of an Ellipse

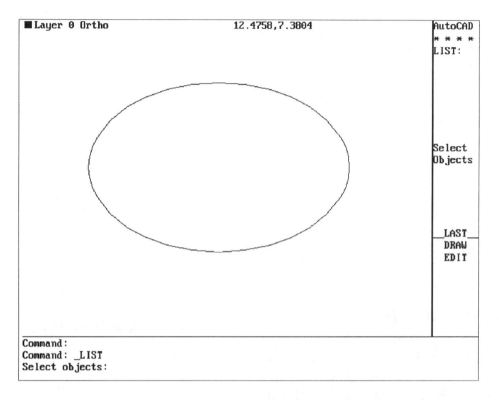

An ellipse is an entity consisting of numerous arcs all converted into one polyline entity. Using the List command on an ellipse displays the following information: The name of the entity being listed, (Polyline); The layer the ellipse was drawn on; Whether the ellipse occupies model or paper space; The specific vertex of the ellipse being listed in absolute coordinates; The origin of the vertex; The starting width of the polyline; The ending width of the polyline; The bulge value of the vertex; The center point of the radius; The radius of the arc segment; The starting angle of the arc segment; The ending angle of the arc segment.

As the ellipse is selected using List, all vertices are listed. When the text screen fills up with information, simply strike the Enter key to list the next series of vertices.

In the illustration above, the List command automatically calculates the area and perimeter of the ellipse.

Using the Explode command on an ellipse breaks the ellipse into individual arc entity segments.

Listing the Properties of an Ellipse using DDMODIFY

```
                          Modify Polyline
  Properties
   [Color...]  ■  BYLAYER           [Layer...] 0
   [Linetype...]  BYLAYER           Thickness: [0.0000        ]

  Polyline Type: 2D polyline        Entity Handle: None
  Vertex Listing   Fit/Smooth       Mesh              Polyline
  [Vertex:1] [Next]  ■ None         M:  □ Closed      ⊠ Closed
                     □ Quadratic    N:  □ Closed      ⊠ LT Gen
   X:  10.0000       □ Cubic        U: [   ]
   Y:  5.0000        □ Bezier       V: [   ]
   Z:  0.0000        □ Curve Fit

                [ OK ]  [ Cancel ]  [ Help... ]
```

Using the DDMODIFY command on an ellipse lists the usual entity properties such as color, linetype, layer name, and entity thickness. Using this command on an ellipse may not allow for dramatic changes such as on arcs and circles. This dialogue box does list the following: The X, Y, and Z coordinates of each vertex; The type of curve fitted to the ellipse; Whether the polyline is closed or open; Whether to generate line type scaling per vertex

The LT Gen option stands for generate a linetype. If LT Gen is unselected in the DDMODIFY dialogue box, or is set to "Off", the linetype is applied to each individual vertex. In the example at the right, with the Ltscale command set to 0.70 only the top and bottom arc segments show with the proper linetype. This is because the other arc segments of the ellipse are too short to support the linetype at the current Ltscale value.

A better result would be to check the LT Gen box in DDMODIFY. This applies the linetype throughout the entire polyline and not per vertex.

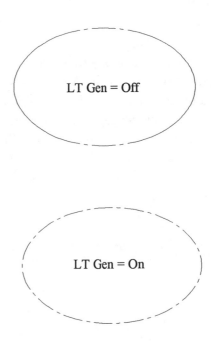

LT Gen = Off

LT Gen = On

Listing the Properties of a Hatch Pattern

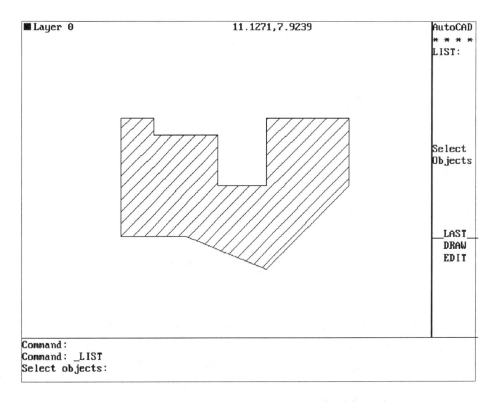

Using the List command on a hatch pattern displays the following information: The name of the entity being listed, (Block Reference); The layer the hatch pattern was drawn on; Whether the hatch pattern occupies model or paper space; The anonymous block name for the hatch pattern, namely "X"; The actual name of the hatch pattern, namely "ANSI31"; The scale of the hatch pattern; The angle of the hatch pattern; The insertion point of the hatch pattern; The X, Y, and Z values for the anonymous block "X"; The rotation angle of the anonymous block "X".

```
BLOCK REFERENCE  Layer: 0
               Space: Model space
    *X
    Hatch pattern ANSI31
       Hatch scale   2.0000
       Hatch angle      0
    at point, X=  0.0000  Y=   0.0000  Z=   0.0000
    X scale factor   1.0000
    Y scale factor   1.0000
rotation angle      0
    Z scale factor   1.0000
```

The Hatch Scale is controlled by the Hatch and Bhatch commands. The X, Y, and Z scale factors relate to the anonymous block name of the hatch pattern, namely "*X". When using a hatch pattern on a boundary area, the user is never prompted for the X, Y, and Z scale factors; only the Hatch Scale factor. Therefore the user should pay attention only to the Hatch Scale factor.

Listing the Properties of a Hatch Pattern using DDMODIFY

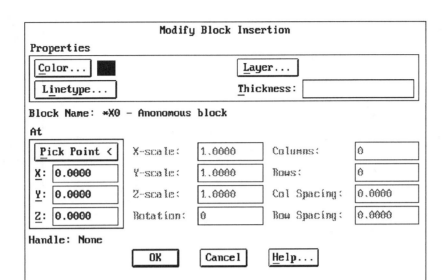

Using this DDMODIFY command dialogue box on a hatch pattern allows the user to dynamically change each of the following items: The insertion point of the block; The X, Y, and Z scale values of the block; The rotation angle of the block.

This dialogue box also displays the possibility of performing multiple inserts of a hatch pattern by allowing the user to change each of the following items similar to that of a block: The number of columns; The number of rows; A value for the spacing inbetween columns; A value for the spacing inbetween rows.

However since it would be impractical to perform multiple insertions on hatch patterns especially as hatch pattern boundaries are seldom alike, this area of the DDMODIFY dialogue box is "greyed out" signifying it cannot be used on this type of entity.

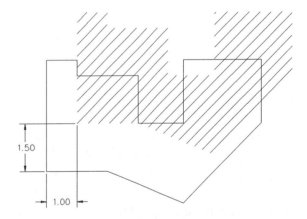

In the same manner, changing the X, Y, and Z insertion values from 0.0000,0.0000,0.0000 to another value performs a shift on the hatch pattern. In the illustration above, the X insertion value was changed to 1.00 and the Y insertion value to 1.50. This again yields impractical results for a hatch pattern.

Listing the Properties of a Line

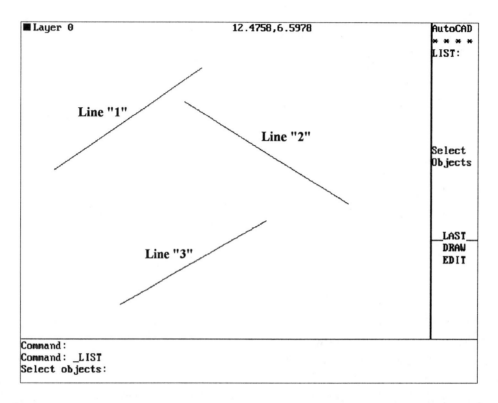

Using the List command on a line entity lists the following information: The specific specific entity being listed (Line); The layer the line segment was drawn on; Whether the line occupies model or paper space; The starting endpoint of the line segment; The ending endpoint of the line segment; The length of the line segment; The angle of the line segment in the X-Y plane; The relative coordinate value from the starting point to the ending point of the line segment.

Study the illustrations at the right and above to isolate the information on the individual line segments. Number of decimal places for the beginning and ending points of the line in addition to the length of the line are governed by the Units command.

```
                    Line "1"

       LINE      Layer: 0
                 Space: Model space
    from point, X=  1.0000  Y=  5.0000  Z=  0.0000
      to point, X=  5.5000  Y=  8.0000  Z=  0.0000
  Length =  5.4003,  Angle in X-Y Plane =    34
         Delta X =  4.5000, Delta Y =   3.0000, Delta Z =   0.0000
```

```
                    Line "2"

       LINE      Layer: 0
                 Space: Model space
    from point, X=  5.0000  Y=  7.0000  Z=  0.0000
      to point, X= 10.0000  Y=  4.0000  Z=  0.0000
  Length =  5.8310,  Angle in X-Y Plane =   329
         Delta X =  5.0000, Delta Y =  -3.0000, Delta Z =   0.0000
```

```
                    Line "3"

       LINE      Layer: 0
                 Space: Model space
    from point, X=  3.0000  Y=  1.0000  Z=  0.0000
      to point, X=  7.5000  Y=  3.5000  Z=  0.0000
  Length =  5.1478,  Angle in X-Y Plane =    29
         Delta X =  4.5000, Delta Y =   2.5000, Delta Z =   0.0000
```

Listing the Properties of a Line using DDMODIFY

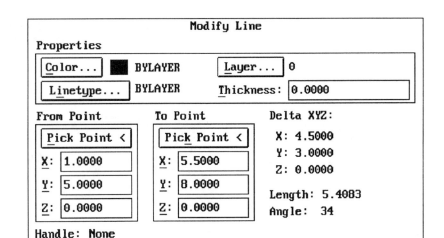

Using the DDMODIFY command on a line segment lists the following information: The X, Y, and Z coordinates of the starting of the line; The X, Y, and Z coordinates of the end of the line; The Delta XYZ coordinate value of the line; The total length of the line segment; The angle the line segment makes in the X-Y plane.

The X, Y, and Z values in the dialogue box above may be changed to affect the beginning or end of the line segment. When any of these values change, the Delta XYZ, Length, and Angle values update themselves to the new values of the line segment.

The first illustration at the right shows how the length and angle are calcualted. Angles will always be calculated in the counterclock-wise direction depending on the current setting in the Units command.

The Delta XY value at the right shows the horizontal and vertical distances from the beginning of the line segment to the end of the line segment.

Line Length and Angle

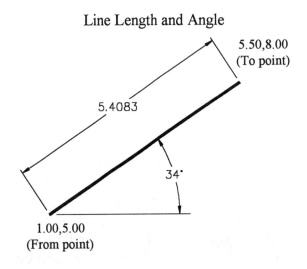

Delta X,Y Values

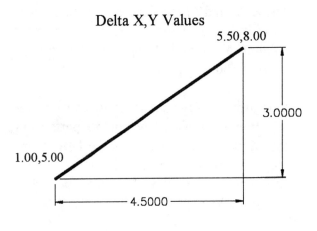

Listing the Properties of a Multiple Block Insertion (MINSERT)

```
■Layer 0                        12.1495,5.0978              AutoCAD
                                                            * * * *
                                                            LIST:


    I      I      I      I      I


                                                            Select
    I      I      I      I      I                           Objects



    I      I      I      I      I                           LAST_
                                                            DRAW
                                                            EDIT

    I      I      I      I      I


Command:
Command: _LIST
Select objects:
```

Using the List command on an block which has been inserted numerous times using the Minsert command provides the user with the following information: The name of the entity being listed, (Block Reference); The layer the block was inserted on; Whether the block occupies model or paper space; The name of the block which in this case is "I-Beam"; The insertion point of the block in X, Y, and Z coordinates; The X, Y, and Z scale factors of the block; The rotation angle of the block; The number of columns along with the spacing; The number of rows and the spacing.

It is always very important to consider the insertion point of the block before performing the Minsert command. For the I-Beam example illustrated above, both the column and row spacing relate directly to the block insertion point located at "A".

```
            BLOCK REFERENCE  Layer: 0
                      Space: Model space
            I-BEAM
            at point, X=   0.8133  Y=   0.9651  Z=   0.0000
             X scale factor    1.0000
             Y scale factor    1.0000
      rotation angle       0
             Z scale factor    1.0000
     # columns 5
column spacing    2.2500
       # rows 4
  row spacing    2.0000
```

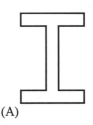

(A)

Listing the Properties of a Multiple Block Insertion using DDMODIFY

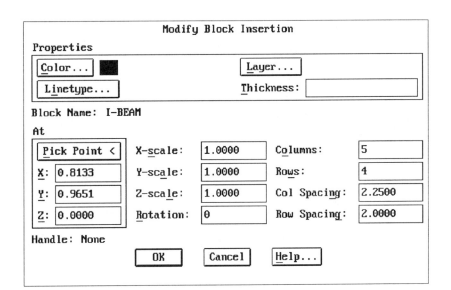

Control of blocks inserted using the Minsert command are enhanced through the use of the DDMODIFY command which displays the dialogue box illustrated above. Using this dialogue box allows the user to dynamically change each of the following items: The insertion point of the block; The X, Y, and Z scale values of the block; The rotation angle of the block.

This dialogue box also controls multiple inserts of a block by allowing the user to change each of the following items: The number of columns; The number of rows; A value for the spacing inbetween columns; A value for the spacing inbetween rows.

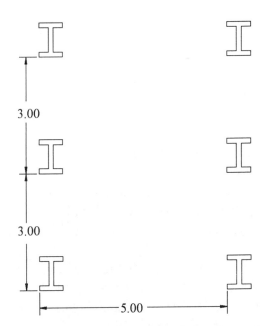

See the illustration above to use DDMODIFY to control multiple block insertions. With the insertion point of the block at the center of the bolt head, 2 columns and 3 rows are specified. Spacing between columns is 5.00 units and between rows is 3.00 units. The results are illustrated at the right.

Of interest is the use of the insertion point as the point of reference where all values for column and row spacing are calculated from. With the lower left corner of the column used as the insertion point, notice the column and row spacing span from the insertion point of one block to the insertion point of the other block.

Listing the Properties of a Point

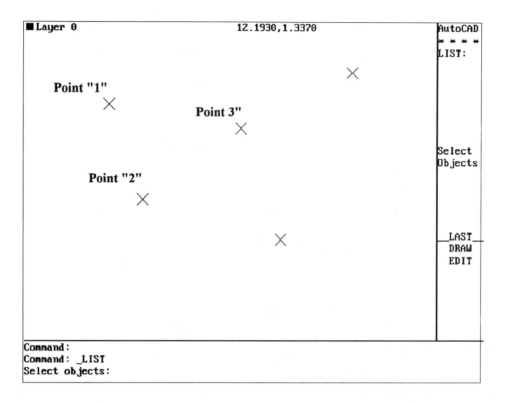

Using the List command on an point entity provides the user with the following information: The name of the entity being listed, (Point); The layer the point was drawn on; Whether the point occupies model or paper space; The current point "Handle"; The coordinates of the point in X, Y, and Z values.

As with all points, the size and appearance is dictated by the system variables PDSIZE and PDMODE. In the example above, a PDMODE of 3 has been specified forming all points in the appearance of an "X".

Study the typical screens at the right and see how they relate to the points in the illustration above.

```
                  Point "1"
   POINT     Layer: 0
             Space: Model space
      Handle = 1
at point, X=   2.5482  Y=   6.9289  Z=   0.0000
```

```
                  Point "2"
   POINT     Layer: 0
             Space: Model space
      Handle = 3
at point, X=   3.5675  Y=   4.1313  Z=   0.0000
```

```
                  Point "3"
   POINT     Layer: 0
             Space: Model space
      Handle = 2
at point, X=   6.5819  Y=   6.2133  Z=   0.0000
```

Listing the Properties of a Point using DDMODIFY

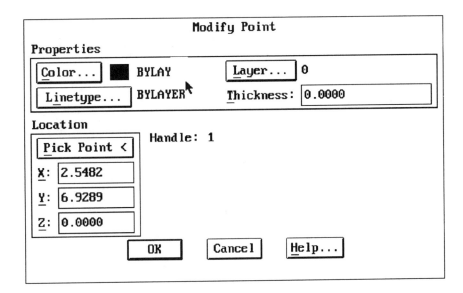

Using the DDMODIFY command on a point lists the usual entity properties such as color, linetype, layer name, and entity thickness. This command also lists the following information: The X, Y, and Z coordinates of the point; The current Handle assigned to the point.

Notice in the above illustration the appearance of the X, Y, and Z coordinate values located in edit boxes. A new value may be entered in one of these boxes which will change the location of the point.

If the absolute coordinates of a point to move are not known, the "Pick Point<" button may be used to locate the point with the current pointing device such as a digitizing puck or mouse. Object snap modes are usually used to locate the point on an entity using the "Pick Point<" button.

Listing the Properties of a Polyline

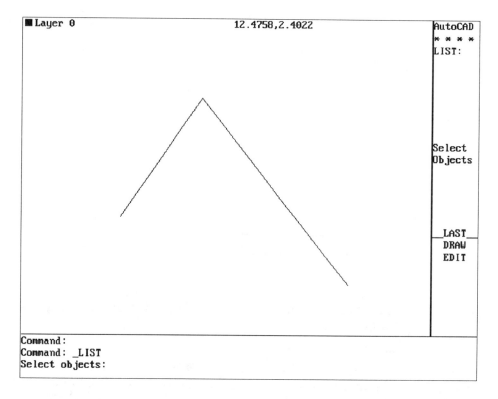

```
Layer 0                    12.4758,2.4022              AutoCAD
                                                       * * * *
                                                       LIST:



                                                       Select
                                                       Objects


                                                       LAST
                                                       DRAW
                                                       EDIT




Command:
Command: _LIST
Select objects:
```

Using the List command on a polyline entity lists the absolute coordinate values of each vertex of the polyline along with the following information: The name of the entity being listed, (Polyline); The layer the polyline was drawn on; Whether the polyline occupies model or paper space; The specific vertex of the polyline being listed in absolute coordinates; The absolute coordinate value at the vertex; The starting width of the vertex; The ending width of the vertex.

Using the Explode command on a polyline separates the polyline into individual line segments.

```
         VERTEX    Layer: 0
                   Space: Model space
         at point, X=   3.0000  Y=   3.5000  Z=   0.0000
starting width    0.0000
  ending width    0.0000

         VERTEX    Layer: 0
                   Space: Model space
         at point, X=   5.5000  Y=   7.0000  Z=   0.0000
starting width    0.0000
  ending width    0.0000

         VERTEX    Layer: 0
                   Space: Model space
         at point, X=  10.0000  Y=   1.5000  Z=   0.0000
starting width    0.0000
  ending width    0.0000

         END SEQUENCE  Layer: 0
                   Space: Model space
     area   14.7500
   length   11.4075
```

Once the last screen is displayed in the illustration above, the List command calculates the area occupied by a closed or open polyline. For an open polyline, the total length of the polyline is given. If the polyline was closed, the permeter of the polyline would be calculated.

Listing the Properties of a Polyline using DDMODIFY

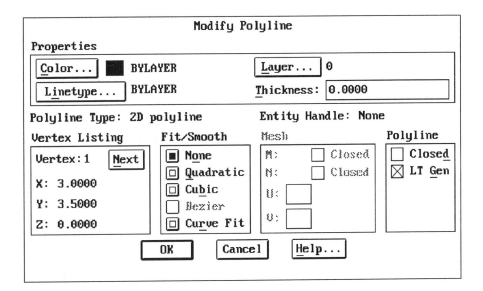

Using the DDMODIFY command on a polyline lists the usual entity properties along with the following information: The X, Y, and Z coordinates of each polyline vertex; The type of curve fitted to the polyline; Whether the polyline is closed or open; Whether to generate line type scaling per vertex.

By default, a polyline has normal curve generation which means is is absent of any curves. Selecting Quadratic, Cubic, and Fit Curve displays the results similar to the illustration at the right.

The LT Gen option stands for generate a linetype. If LT Gen is unselected in the DDMODIFY dialogue box, or is set to "Off", the linetype is applied to each individual vertex. In the example at the right, with the Ltscale command set to 1.10, one leg of the polyline has a single center line while the other leg has three center lines. Checking the LT Gen box above in DDMODIFY turns on LT Gen with the results at the far right. Here, the first leg of the polyline has two center lines that continue into the second leg.

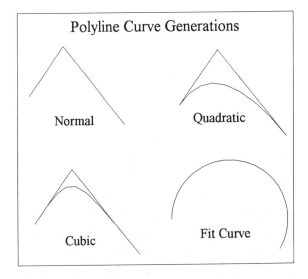

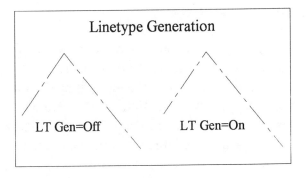

Listing the Properties of a Polygon

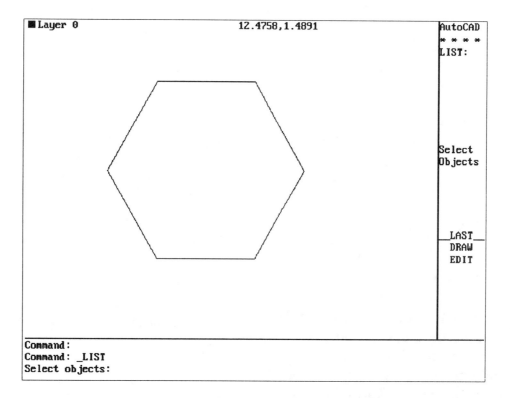

Using the List command on a polygon entity lists the absolute coordinate values of each vertex of the polygon along with the following information: The name of the entity being listed, (Polyline); The layer the polygon was drawn on; Whether the polygon occupies model or paper space; The specific vertex of the polygon being listed in absolute coordinates; The absolute coordinate value at the vertex; The starting width of the vertex; The ending width of the vertex.

Depending on the number of vertices, sometimes this information will not fit on the display screen all at once. Once the screen is filled with information, strike the Enter key to continue listing more vertex information. Once the last screen is displayed, the List command calculates the area occupied by polygon.

```
        VERTEX    Layer: 0
                  Space: Model space
        at point, X=  2.4783  Y=  4.9325  Z=  0.0000
starting width   0.0000
 ending width    0.0000

        VERTEX    Layer: 0
                  Space: Model space
        at point, X=  3.9783  Y=  2.3345  Z=  0.0000
starting width   0.0000
 ending width    0.0000

        VERTEX    Layer: 0
                  Space: Model space
        at point, X=  6.9783  Y=  2.3345  Z=  0.0000
starting width   0.0000
 ending width    0.0000

        END SEQUENCE Layer: 0
                  Space: Model space
     area   23.3827
perimeter   18.0000
```

Listing the Properties of a Polygon using DDMODIFY

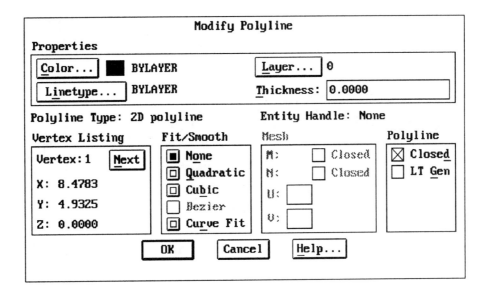

Using the DDMODIFY command on a polygon lists the usual entity properties such as color, linetype, layer name, and entity thickness. Using this command lists the following information: The X, Y, and Z coordinates of each polygon vertex; The type of curve fitted to the polygon; Whether the polygon is closed or open; Whether to generate line type scaling per vertex.

If LT Gen is unselected in the DDMODIFY dialog box, or is set to "Off", the linetype is applied to each individual vertex. In the example at the right, with the Ltscale command set to 1.10 each leg of the polyline has a single center line. Checking the LT Gen box above in DDMODIFY turns on LT Gen with the results at the far right where the linetype is distributed throughout the entire polyline.

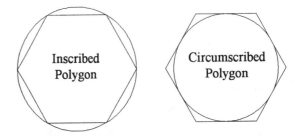

Illustrated above are examples of an inscribed polygon, (inside of a circle) and a circumscribed polygon, (around a circle).

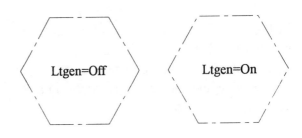

Listing the Properties of Text

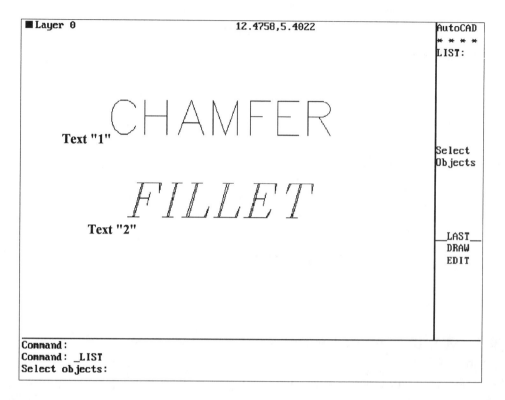

Using the List command on text displays the following information: The name of the entity being listed, (Text); The layer the entity was drawn on; Whether the entity occupies model or paper space; The text style the entity was drawn in; The font file the entity was drawn in; The starting point of the text; The text height; The actual text that was entered from the keyboard; The text rotation angle; The current width factor of the text; The current obliquing angle for the text; The type of text generation.

In the examples above and to the right, the width scale factor, obliquing angle, and generation of text are all controlled by the Style command when the text font was originally associated with a text style.

```
                    Text "1"

          TEXT       Layer: 0
                     Space: Model space
      Style = STANDARD    Font file = SIMPLEX
    center point, X=   6.0000  Y=   6.0000  Z=   0.0000
    height    1.0000
       text CHAMFER
    rotation angle      0
       width scale factor      1.0000
  obliquing angle      0
  generation normal
```

```
                    Text "2"

          TEXT       Layer: 0
                     Space: Model space
      Style = ITALICC    Font file = ITALICC
     start point, X=   3.1711  Y=   3.5867  Z=   0.0000
    height    1.0000
       text FILLET
    rotation angle      0
       width scale factor      1.0000
  obliquing angle      0
  generation normal
```

Listing the Properties of Text using DDMODIFY

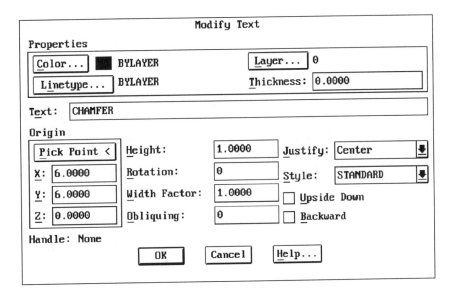

Using the DDMODIFY command on a text entity provides for superior control of text. All listings appearing in a box all may be dynamically changed to affect the final form of the text entity. In addition to the usual entity properties that may be changed such as color, linetype, layer name, and entity thickness, the following may be changed using this dialogue box: The actual text may be edited in a way similar to the DDEDIT command; Selecting a new text origin point; Entering a new text height; Entering a new rotation angle for the text; Entering a new width factor for the Text; Entering an obliquing angle to make text inclined; Selecting a new justification position for the text; Selecting a new text style.

Keep in mind that all of the above changes only apply to the text entity selected and does not globally affect all text entities.

Selecting the current text style opens another dialogue box illustrated at the right. Use this dialogue box to select additional text styles. These styles are defined using the Style command.

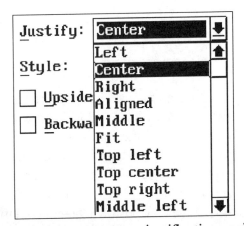

Selecting the current text justification position opens up an additional dialogue box illustrated above. All valid text justification positions appear allowing the user to scroll up or down to select a new justification position.

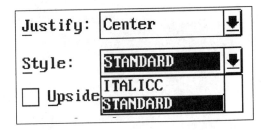

Listing the Properties of a Viewport

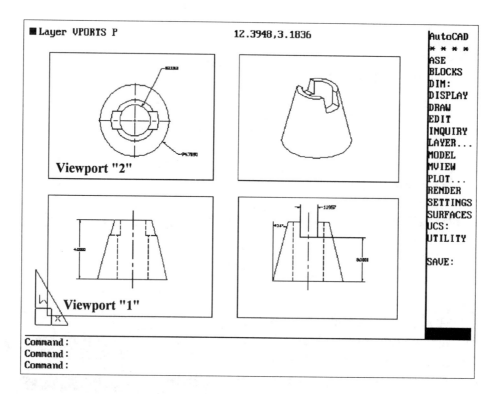

Using the List command on an point entity provides the user with the following information: The name of the entity being listed, (Viewport); The layer the viewport was drawn on; The space occupied by the viewport, namely Paper space; The current viewport "Handel"; The status of the viewport, whether On or Off; The scale of the viewport relative to Paper space; The center point of the viewport in X, Y, and Z coordinate values; The width of the viewport; The height of the viewport.

```
                  Viewport "1"

       VIEWPORT  Layer: VPORTS
                 Space: Paper space
          Handle = D0
                 Status: On and Active
                 Scale relative to Paper space:    0.5000xp
       center point, X=   2.9313  Y=   2.1957  Z=  0.0000
        width    5.7104
        height   4.0652
```

```
                  Viewport "2"

       VIEWPORT  Layer: VPORTS
                 Space: Paper space
          Handle = D1
                 Status: On and Active
                 Scale relative to Paper space:    0.5000xp
       center point, X=   2.9402  Y=   6.8949  Z=  0.0000
        width    5.7225
        height   4.2367
```

Listing the Properties of a Viewport using DDMODIFY

Viewport "1"

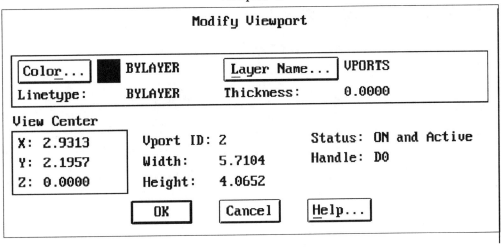

Viewport "2"

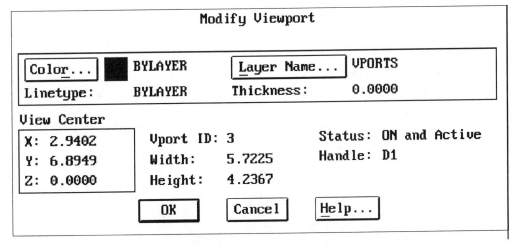

Using the DDMODIFY command on a viewport displays the information in the illustration above. A viewports linetype and thickness cannot be changed. As with using DDMODIFY on an associative dimension, this command displays information identical to using the List command. As the View Center is listed, it may not be changed.

Using the Status Command

```
1736 entities in C:\Q21\SAMPLE\TOOLPOST
Model space limits are X:    0.0000   Y:    0.0000   (Off)
                       X:   18.0000   Y:   14.0000
Model space uses       X:    3.7500   Y:    4.7500
                       X:   17.2500   Y:   13.2500
Display shows          X:    3.7500   Y:    4.7500
                       X:   17.3859   Y:   14.6259
Insertion base is      X:    0.0000   Y:    0.0000   Z:    0.0000
Snap resolution is     X:    1.0000   Y:    1.0000
Grid spacing is        X:    0.0000   Y:    0.0000

Current space:      Model space
Current layer:      0
Current color:      BYLAYER -- 7 (white)
Current linetype:   CONTINUOUS
Current elevation:  0.0000  thickness:   0.0000
Fill on  Grid off  Ortho off  Qtext off  Snap off  Tablet off
Object snap modes:   None
Free disk: 8177664 bytes
Virtual memory allocated to program: 3624K
Amount of program in physical memory/Total (virtual) program size: 67%
-- Press RETURN for more --
Total conventional memory: 292K     Total extended memory: 7424K
Swap file size: 388K bytes
```

Using the Status Command

Once inside of a large drawing, it becomes difficult sometimes to keep track of various settings that have been changed from their default values to different values required by the drawing. To obtain a listing of these important settings contained in the current drawing file, use the Status command. Once the Status command is invoked, the graphics screen changes to a text screen displaying all of the information illustrated at the right. The following information is supplied by the Status command:
- The number of entities contained in the current drawing file.
- The current Model space limits set by the Limits command.
- The area used by Model space.

If the current work environment is Paper space, the current limits and area used will be listed as Paper space limits. The following additional information is supplied by the Status command:
- The coordinates of the current drawing display.
- The current insertion base point of the drawing.
- The current snap resolution value.
- The current grid spacing.
- The current space occupied by the drawing (Model or Paper).
- The current layer of the drawing.

- The current color.
- The current linetype.
- The current elevation and thickness of the drawing.
- Whether the following modes are On of Off:
    - Fill
    - Grid
    - Ortho
    - Qtext
    - Snap
    - Tablet
- The current running Object snap modes currently activated.

The remaining lines appearing after the Object snap modes deal with the operating system of the central processing unit of the computer used to generate the current drawing. One of these parameters, Free disk, refers to the space left on the hard disk drive that contains the AutoCAD drawing. If disk space runs out, AutoCAD terminates; however not before giving the operator the opportunity to save their work.

The following information is **not** provided at the bottom of the display when using the Status command in the Windows version of AutoCAD: Virtual memory allocated to program; Amount of program in physical memory/Total (virtual) program size; Total conventional memory; Total extended memory; Swap file size.

Using the Time Command

```
Command: _TIME
Current time:           22 Jun 1992 at 21:25:58.340
Drawing created:        21 Aug 1990 at 15:24:58.640
Drawing last updated:   14 Apr 1992 at 20:45:47.990
Time in drawing editor: 0 days 00:25:50.700
Elapsed timer:          0 days 00:25:50.700
Next automatic save in: 0 days 01:56:56.160
Timer on.
Display/ON/OFF/Reset:
```

The Time command provides the operator with the following information: The current date and time; The date and time the drawing was created; The last time the drawing was updated; The total time spent editing the drawing so far; The total time spent while in AutoCAD, not necessarily in a particular drawing; The current automatic save time interval.

Current Time:

All dates and times are set by the DOS Date and Time commands. If using the Time command, it is important to make sure these DOS commands are properly set in order to display the desired results.

Drawing Created:

This date and time value is set whenever using the New command for creating a new drawing file. This value is also set to the current date and time whenever a Wblock is created or a drawing is saved under a different name using the Save command.

Drawing Last Updated:

This data consists of the date and time the current drawing was last updated. This value updates itself whenever using the Save command or the End command.

Time in Drawing Editor:

This represents the total time spent editing the drawing. The timer is always updating itself and cannot be reset to a new or different value.

Elapsed Timer:

This timer runs while AutoCAD is in operation and can be turned on or off or reset by the user.

Next Automatic Save In:

This timer displays when the next automatic save will occur. This value is controlled by the system variable "SAVETIME". If this system variable is set to zero, the automatic save utility is disabled. If the timer is set to a nonzero value, the timer displays when the next automatic save will take place. The increment for automatic saving is in minutes.

UNIT 4

Inquiry Command Tutorial Exercises

This unit consists of three tutorials designed to give the user practice on constructing an object and then answering a series of questions through the use of Inquiry commands. All tutorial exercises begin with a step-by-step format for constructing an object. As this represents one method for building an object, numerous other methods could be used.

Once an object has been constructed, the tutorial continues by describing how various Inquiry commands are used to answer various questions about the object. Typical questions ask the individual to find area, perimeter, distance, delta diatance, point identification, and the angle formed in the XY plane.

Proceed through the tutorial exercises at your own pace. Similar questions have already been asked in the Pre-Test material; more questions are asked in two Post-Tests that follow.

Tutorial Exercise #1
Extrude.Dwg

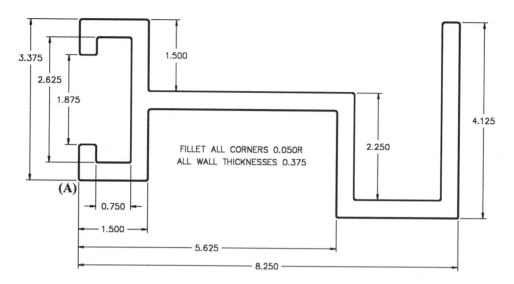

FILLET ALL CORNERS 0.050R
ALL WALL THICKNESSES 0.375

Purpose:

This tutorial is designed to show the user various methods in constructing the extruded pattern above. The surface area of the extrusion will also be found using the Area command.

System Settings:

Keep the default drawing limits at 0.0000,0.0000 for the lower left corner and 12.0000,9.0000 for the upper right corner. Use the Units command and change the number of decimal places past the zero from 4 units to 3 units.

Layers:

No special layers need be created for this drawing although it is always considered good practice to create and draw on a separate layer for all object lines.

Suggested Commands:

Begin drawing the extrusion with point "A" illustrated above at absolute coordinate 2.000,3.000. Use either of the following methods to construct the extrusion:

1. Using a series of absolute, relative, and polar coordinates to construct the profile of the extrusion.
2. Constructing a few lines; then using the Offset command followed by the Trim command to construct the extrusion profile.

The Fillet command is used to create the 0.050 radius rounds at all corners of the extrusion. Before calculating the area of the extrusion, convert and join all entities into one single polyline. This will allow the Area command to be used in a more productive way.

Dimensioning:

This drawing does not need to be dimensioned in order to answer any Inquiry command question.

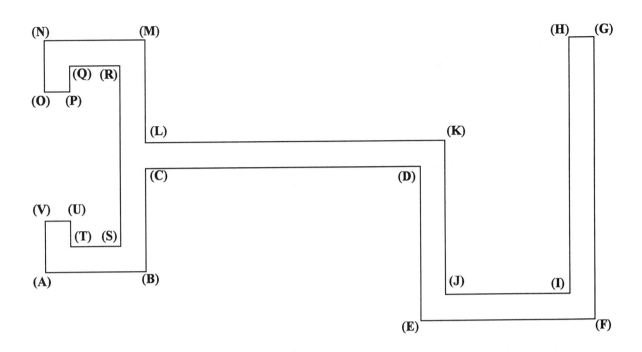

Step #1

One method of constructing the extrusion above is to use the measurements on the previous page to calculate a series of polar coordinate system distances.

Command: **Line**
From point: **2.000,3.000** *(Starting at "A")*
To point: **@1.500<0** *(To "B")*
To point: **@1.500<90** *(To "C")*
To point: **@4.125<0** *(To "D")*
To point: **@2.250<270** *(To "E")*
To point: **@2.625<0** *(To "F")*
To point: **@4.125<90** *(To "G")*
To point: **@0.375<180** *(To "H")*
To point: **@3.750<270** *(To "I")*
To point: **@1.875<180** *(To "J")*
To point: **@2.250<90** *(To "K")*

To point: **@4.500<180** *(To "L")*
To point: **@1.500<90** *(To "M")*
To point: **@1.500<180** *(To "N")*
To point: **@0.750<270** *(To "O")*
To point: **@0.375<0** *(To "P")*
To point: **@0.375<90** *(To "Q")*
To point: **@0.750<0** *(To "R")*
To point: **@2.625<270** *(To "S")*
To point: **@0.750<180** *(To "T")*
To point: **@0.375<90** *(To "U")*
To point: **@0.375<180** *(To "V")*
To point: **Close** *(Back to "A")*

Step #2

All corners come together at 90 degree intersections. From the original dimensions of the extrusion, a note calls out that all corners be rounded off with a 0.500 radius. This is easily accomplished using the Fillet command. However since so many corners need to be filleted, the risk is high of forgetting to fillet one or more corners. All corners may be filleted at one time only if the entire extrusion consists of one polyline. First the Pedit command will be used to perform this conversion followed by the Fillet command.

Command: **Pedit**
Select polyline: *(Select the entity labeled "A")*
Entity selected is not a polyline.
Do you want to turn it into one? <Y>: *(Strike Enter to continue)*
Close/Join/Width/Edit vertex/Fit/Spline/ Decurve/Ltype gen/Undo/eXit <X>: **Join**
Select objects: **W** *(To window in all entities)*
First corner: **0.000,0.000**
Other corner: **12.000,9.000**
Select objects: *(Strike Enter to continue)*
21 segments added to polyline
Open/Join/Width/Edit vertex/Fit/Spline/ Decurve/Ltype gen/Undo/eXit <X>: *(Strike Enter to exit this command)*

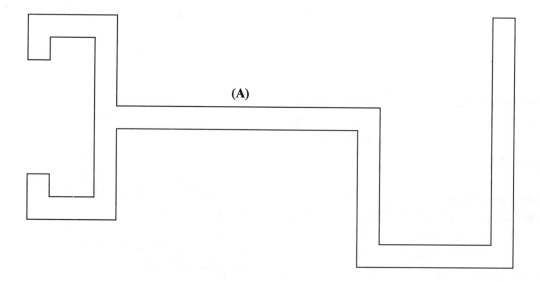

(A)

Step #3

With the entire extrusion converted into a polyline, use the Fillet command, set a radius of 0.050, and use the polyline option of the Fillet command to fillet all corners of the extrusion at once.

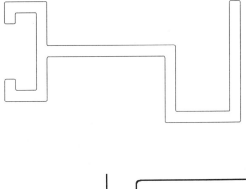

Command: **Fillet**
Polyline/Radius/<Select first object>: **Radius**
Enter fillet radius <0.0000>: **0.050**

Command: **Fillet**
Polyline/Radius/<Select first object>: **Polyline**
Select 2D polyline: *(Select the polyline at the right)*
22 lines were filleted

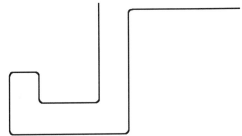

Checking the Accuracy of Extrude.Dwg

Once the extrusion has been constructed, answer the question below to determine the accuracy of the drawing.

Question #1
What is the total surface area of the Extrusion?

Use the Area command to calculate the surface area of the extrusion. This is easily accomplished since the extrusion has already been converted into a polyline.

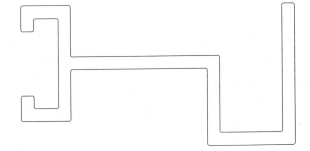

Command: **Area**
<First point>/Entity/Add/Subtract: **Entity**
Select circle or polyline: *(Select any part of the extrusion)*
Area = 7.170,　　　Perimeter = 38.528

Total surface area of the Extrusion is 7.170

Notes

Tutorial Exercise #2
C-Lever.Dwg

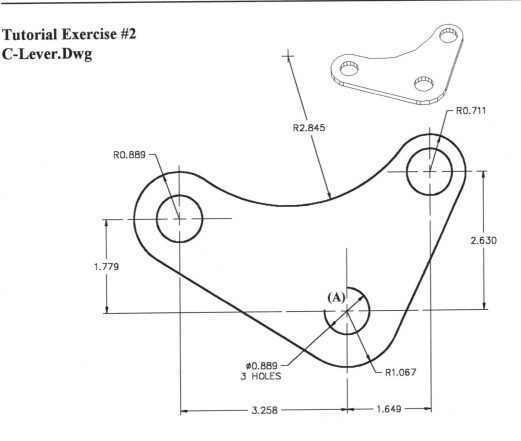

Purpose:

This tutorial is designed to show the user various methods in constructing the C-Lever object above. Numerous questions will be asked about the object requiring the use of a majority of Inquiry commands.

System Settings:

Keep the default drawing limits at 0.0000,0.0000 for the lower left corner and 12.0000,9.0000 for the upper right corner. Use the Units command and change the number of decimal places past the zero from 4 units to 3 units.

Layers:

No special layers need be created for this drawing although it is always considered good practice to create and draw on a separate layer for all object lines.

Suggested Commands:

Begin drawing the C-Lever with point "A" illustrated above at absolute coordinate 7.000,3.375. Begin laying out all circles. Then draw tangent lines and arcs. Use the Trim command to clean up unnecessary entities. To prepare to answer the Area command question, convert the profile of the C-Lever into a polyline using the Pedit command. Other question pertaining to distances, angles, and point identifications follow.

Dimensioning:

This drawing does not need to be dimensioned in order to answer any Inquiry command question.

Step #1

Construct one circle of 0.889 diameter with the center of the circle at absolute coordinate 7.000,3.375. Construct the remaining circles of the same diameter by using the Copy command with the multiple option. Use of the "@" symbol for the base point in the copy command identifies the last known point which in this case is the center of the first circle drawn at coordinate 7.000,3.375

Command: **Circle**
3P/2P/TTR/<Center point>: **7.000,3.375**
Diameter/<Radius>: **Diameter**
Diameter: **0.889**

Command: **Copy**
Select objects: **Last**
Select objects: *(Strike Enter to continue)*
<Base point or displacement>/Multiple: **Multiple**
Base point: **@**
Second point of displacement: **@1.649,2.630**
Second point of displacement: **@-3.258,1.779**
Second point of displacement: *(Strike Enter to exit this command)*

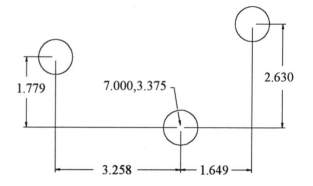

Step #2

Construct three more circles. Eventhough these entities actually represent arcs, circles will be drawn now and trimmed later to form the arcs.

Command: **Circle**
3P/2P/TTR/<Center point>: **Cen**
of *(Select the edge of circle "A")*
Diameter/<Radius>: **1.067**

Command: **Circle**
3P/2P/TTR/<Center point>: **Cen**
of *(Select the edge of circle "B")*
Diameter/<Radius>: **0.889**

Command: **Circle**
3P/2P/TTR/<Center point>: **Cen**
of *(Select the edge of circle "C")*
Diameter/<Radius>: **0.711**

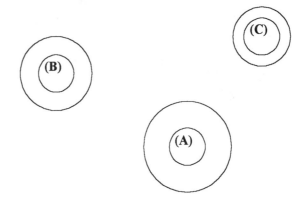

Step #3

Construct lines tangent to the three outer circles illustrated at the right.

Command: **Line**
From point: **Tan**
to *(Select the outer circle near "A")*
To point: **Tan**
to *(Select the outer circle near "B")*
To point: *(Strike Enter to exit this command)*

Command: **Line**
From point: **Tan**
to *(Select the outer circle near "C")*
To point: **Tan**
to *(Select the outer circle near "D")*
To point: *(Strike Enter to exit this command)*

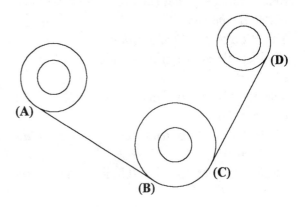

Step #4

Construct a circle tangent to the two circles
illustrated at the right using the Circle command with the Tangent-Tangent-Radius option (TTR).

Command: **Circle**
3P/2P/TTR/<Center point>: **TTR**
Enter Tangent spec: *(Select the outer circle
near "A")*
Enter second Tangent spec: *(Select the outer
circle near "B")*
Radius: **2.845**

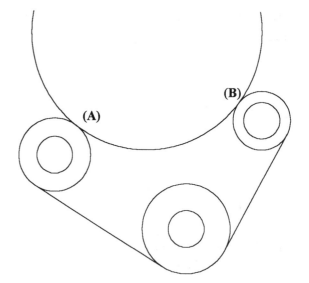

Step #5

Use the Trim command to clean up and form
the finished object. Select all of the entities
represented by dashed lines as cutting edges.
Follow the prompts below for selecting the
entities to trim.

Command: **Trim**
Select cutting edge(s)...
Select objects: *(Select all dashed entities illustrated at the right)*
Select objects: *(Strike Enter to continue)*
<Select object to trim>/Undo: *(Select the
circle at "A")*
<Select object to trim>/Undo: *(Select the
circle at "B")*
<Select object to trim>/Undo: *(Select the
circle at "C")*
<Select object to trim>/Undo: *(Select the
circle at "D")*
<Select object to trim>/Undo: *(Strike Enter to
exit this command)*

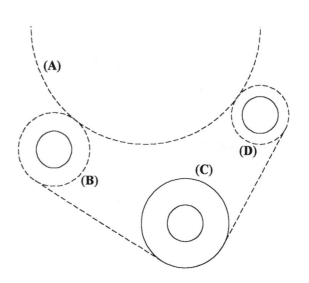

Checking the Accuracy of C-Lever.Dwg

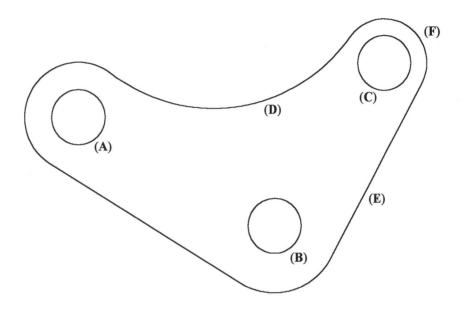

Once the C-Lever has been constructed, answer the questions below to determine the accuracy of the drawing. Use the illustration above to assist in answering the questions.

1. The total area of the C-Lever with all three holes removed is_____

2. The total distance from the center of circle "A" to the center of circle "B" is_____

3. The angle formed in the X-Y plane from the center of circle "C" to the center of circle "B" is

_____

4. The delta X-Y distance from the center of circle "C" to the center of circle "A" is_____

5. The absolute coordinate value of the center of arc "D" is_____

6. The total length of line "E" is_____

7. The total length of arc "F" is_____

A solution for each question follows complete with the method used to arive at the answer. Apply these methods to any type of drawing with similar needs.

Question #1
The total area of the C-Lever with all three holes removed is ?

Refer to the illustration below and the text at the right for the correct procedure used in determining the area of the C-Lever.

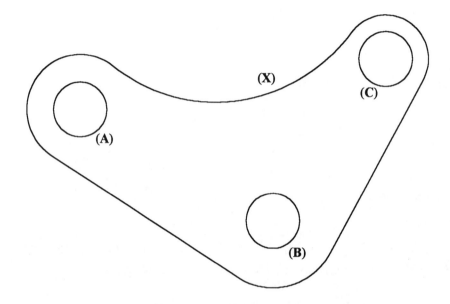

Question #1
The total area of the C-Lever with all three holes removed is ?

The Area command will be used to first calculate the total area of the object and then subtract all three holes. However before using the Area command, all entities representing the outline of the object must be converted into a polyline. The Pedit command with the Join option is used to best accomplish this. Use the illustration of the C-Lever at the right to guide you in the use of the Pedit and Area commands.

Command: **Pedit**
Select polyline: *(Select the entity labeled "X")*
Entity selected is not a polyline.
Do you want to turn it into one? <Y>: *(Strike Enter to continue)*
Close/Join/Width/Edit vertex/Fit/Spline/Decurve/Ltype gen/Undo/eXit <X>: **Join**
Select objects: W
First corner: **0.000,0.000**
Other corner: **12.000,9.000**
Select objects: *(Strike Enter to continue)*
5 segments added to polyline
Open/Join/Width/Edit vertex/Fit/Spline/Decurve/Ltype gen/Undo/eXit <X>: *(Strike Enter to exit this command)*

All outer entities now consist as one polyline entity. Notice that when selecting all entities including the three circles, only the entities connected formed the polyline. Since the circles were independent of the outer entities, they were not included in the creation of the polyline. An alternative method of selecting entities to join into a polyline is to select each entity individually.

Now the Area command may be successfully used to calculate the area of the object with the holes removed.

Command: **Area**
<First point>/Entity/Add/Subtract: **Add**
<First point>/Entity/Subtract: **Entity**
(ADD mode) Select circle or polyline: *(Select the edge of the object near "X")*
Area = 15.611,   Perimeter = 17.771
Total area = 15.611
(ADD mode) Select circle or polyline: *(Strike Enter to continue)*
<First point>/Entity/Subtract: **Subtract**
<First point>/Entity/Add: **Entity**
(SUBTRACT mode) Select circle or polyline: *(Select circle"A")*
Area = 0.621, Circumference = 2.793
Total area = 14.991
(SUBTRACT mode) Select circle or polyline: *(Select circle"B")*
Area = 0.621, Circumference = 2.793
Total area = 14.370
(SUBTRACT mode) Select circle or polyline: *(Select circle"C")*
Area = 0.621, Circumference = 2.793
Total area = 13.749
(SUBTRACT mode) Select circle or polyline: *(Strike Enter to continue)*
<First point>/Entity/Add: *(Strike Enter to exit this command)*

The total area of the C-Lever with all three holes removed is 13.749

Question #2
The total distance from the center of circle "A" to the center of circle "B" is ?

Use the Dist (Distance) command to calculate the distance from the center of circle "A" to the center of circle "B". Be sure to use the Osnap-Center mode. Notice that additional information is given when using the Dist command. For the purpose of this question, we will only be looking for the distance.

Command: **Dist**
First point: **Cen**
of *(Select the edge of circle "A")*
Second point: **Cen**
of *(Select the edge of circle "B")*
Distance = 3.712

The total distance from the center of circle "A" to the center of circle "B" is 3.712

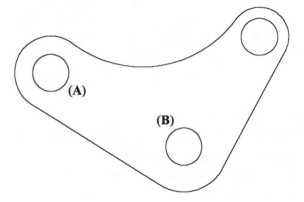

Question #3
The angle formed in the X-Y plane from the center of circle "C" to the center of circle "B" is ?

Use the Dist (Distance) command to calculate the angle from the center of circle "C" to the center of circle "B". Be sure to use the Osnap-Center mode. Notice that additional information is given when using the Dist command. For the purpose of this question, we will only be looking for the angle in the X-Y plane.

Command: **Dist**
First point: **Cen**
of *(Select the edge of circle "C")*
Second point: **Cen**
of *(Select the edge of circle "B")*
Angle in X-Y Plane = 238

The angle formed in the X-Y plane from the center of circle "C" to the center of circle "B" is 238 degrees

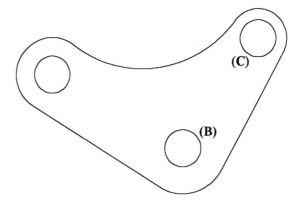

Question #4
The delta X-Y distance from the center of circle "C" to the center of circle "A" is ?

Use the Dist (Distance) command to calculate the delta X-Y distance from the center of circle "C" to the center of circle "A". Be sure to use the Osnap-Center mode. Notice that additional information is given when using the Dist command. For the purpose of this question, we will only be looking for the delta X-Y distance. The Dist command will display the relative X, Y, and Z distances. Since this is a 2 dimensional problem, only the X and Y values will be used.

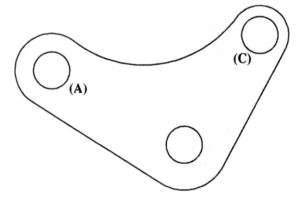

Command: **Dist**
First point: **Cen**
of *(Select the edge of circle "C")*
Second point: **Cen**
of *(Select the edge of circle "A")*
Delta X = -4.907,   Delta Y = -0.851,
Delta Z = 0.000

The delta X-Y distance from the center of circle "C" to the center of circle "A" is
-4.907, -0.851

Question #5
The absolute coordinate value of the center of arc "D" is ?

The ID command is used to get the current absolute coordinate information on a desired point. This command will display the X, Y, and Z coordinate values. Since this is a 2 dimensional problem, only the X and Y values will be used.

Command: **ID**
Point: **Cen**
of *(Select the edge of arc "D")*
X = 5.869, Y = 8.223, Z = 0.000

The absolute coordinate value of the center of arc "D" is 5.869,8.223

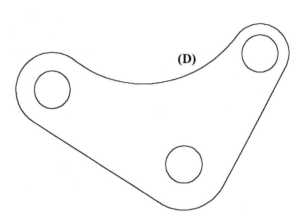

Question #6
The total length of line "E" is ?

Use the Dist (Distance) command to find the total length of line "E". Be sure to use the Osnap-Endpoint mode. Notice that additional information is given when using the Dist command. For the purpose of this question, we will only be looking for the distance.

Command: **Dist**
First point: **Endp**
of *(Select the endpoint of the line at "X")*
Second point: **Endp**
of *(Select the endpoint of the line at "Y")*
Distance = 3.084

The total length of line "E" is 3.084

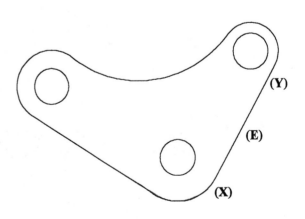

Question #7
The total length of arc "F" is ?

The List command is used to calculate the lengths of arcs; however a little preparation is needed before preforming this operation. If arc "F" is selected at the right, notice the entire outline is selected since it is a polyline. Use the Explode command to break the outline back into individual entities. Again use the List command on arc "F". The following information is displayed: Entity name, Layer entity is found on, Model Space or Paper Space, Center point of the arc, Radius of the arc, Starting and ending angles of the arc. However, the length of the arc is not given. Convert the arc into a polyline using the Pedit command and then use List. Along with the vertices of the polyarc, the area and length are given.

Command: **Explode**
Select objects: *(Select the dashed polyline anywhere)*
Select objects: *(Strike Enter to execute this command)*

Command: **Pedit**
Select polyline: *(Select arc "F")*
Entity selected is not a polyline.
Do you want to turn it into one? <Y>: *(Strike Enter to continue)*
Close/Join/Width/Edit vertex/Fit/Spline/ Decurve/Ltype gen/Undo/eXit <X>: *(Strike Enter to exit this command)*

Command: **List**
Select objects: *(Select arc "F")*
Select objects: *(Strike Enter to continue)*
Length = 2.071

The total length of arc "F" is 2.071

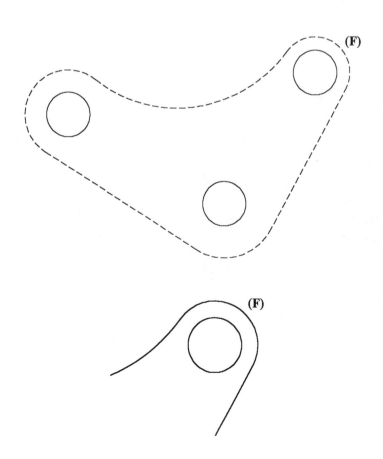

Tutorial Exercise #3
Fitting.Dwg

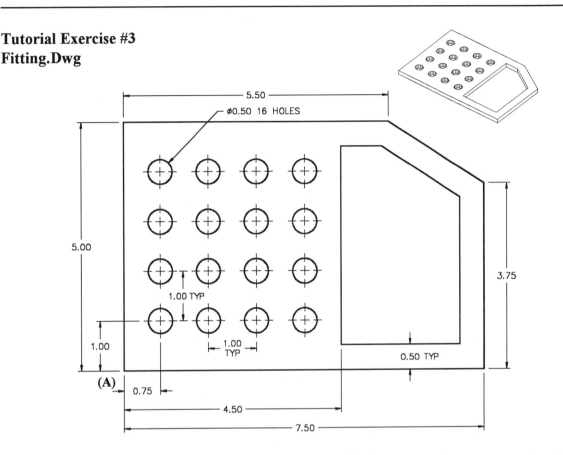

Purpose:

This tutorial is designed to show the user various methods in constructing the Fitting object above. Numerous questions will be asked about the object requiring the use of a majority of Inquiry commands.

System Settings:

Use the Units command and change the number of decimal places past the zero from 4 units to 2 units. Keep the default drawing limits at 0.00,0.00 for the lower left corner and 12.00,9.00 for the upper right corner.

Layers:

No special layers need be created for this drawing although it is always considered good practice to create and draw on a separate layer for all object lines.

Suggested Commands:

Begin drawing the Fitting with point "A" illustrated above at absolute coordinate 2.24,1.91. Begin by laying out the profile of the Fitting. Locate one circle and use the Array command to produce 4 rows and columns of the circle. Use a series of Offset, Trim, and Fillet commands to construct the 5 sided figure on the inside of the Fitting profile. To prepare for the Area question, convert the outer and inner profiles into a polyline using the Pedit command. Other question pertaining to distances, angles, and point identifications follow.

Dimensioning:

This drawing does not need to be dimensioned in order to answer any Inquiry command question.

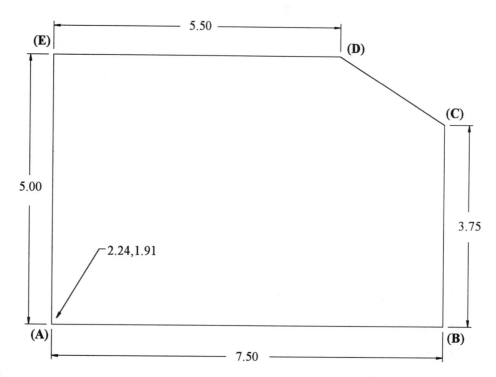

Step #1

Use the Line command to construct the outline of the Fitting using a combination of absolute, polar, and relative coordinates. Begin the lower left corner of the Fitting at absolute coordinate 2.24,1.91. Do not use the Close option of the Line command for constructing the last side of the Fitting.

Command: **Line**
From point: **2.24,1.91** *(Starting at "A")*
To point: **@7.50<0** *(To "B")*
To point: **@3.75<90** *(To "C")*
To point: **@-2.00,1.25** *(To "D")*
To point: **@5.50<180** *(To "E")*
To point: **@5.00<270** *(Back to "A")*
To point: *(Strike Enter to exit this command)*

Step #2

Construct a circle of 0.50 diameter using the Circle command. Since the last known point is at "A" from use of the previous Line command, this point is referenced using the "@" symbol followed by a coordinate value for the center point. This identifies the center of the circle from the last known point.

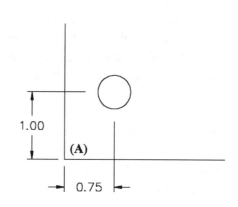

Command: **Circle**
3P/2P/TTR/<Center point>: **@0.75,1.00**
Diameter/<Radius>: **Diameter**
Diameter: **0.50**

Step #3

Create multiple copies in a rectangular pattern of the last circle by using the Array command. There are 4 rows and columns each with a spacing of 1.00 units from the center of one circle to the center of the other. Since the array is to the right and up from the existing circle, all 1.00 spacing units are positive.

Command: **Array**
Select objects: **Last** *(This should select the circle)*
Select objects: *(Strike Enter to continue)*
Rectangular or Polar array (R/P): **Rectangular**
Number of rows (---) <1>: **4**
Number of columns (||||) <1>: **4**
Unit cell or distance between rows (---): **1.00**
Distance between columns (||||): **1.00**

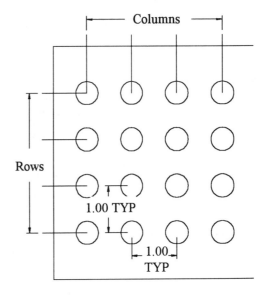

Step #4

Use the Offset command copy the line at "A" parallel at a distance of 4.50 in the direction of "B".

Command: **Offset**
Offset distance or Through <Through>: **4.50**
Select object to offset: *(Select line "A")*
Side to offset? *(Pick a point anywhere near "B")*
Select object to offset: *(Strike Enter to exit this command)*

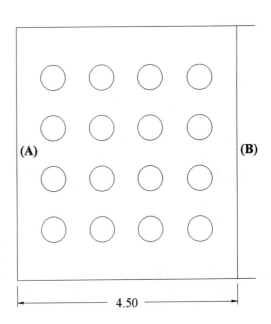

Step #5

Use the Offset command copy the lines at "A", "B", "C", and "D" parallel at a distance of 0.50 in the direction of "E".

Command: **Offset**
Offset distance or Through <Through>: **0.50**
Select object to offset: *(Select line "A")*
Side to offset? *(Pick a point anywhere near "E")*
Select object to offset: *(Select line "B")*
Side to offset? *(Pick a point anywhere near "E")*
Select object to offset: *(Select line "C")*
Side to offset? *(Pick a point anywhere near "E")*
Select object to offset: *(Select line "D")*
Side to offset? *(Pick a point anywhere near "E")*
Select object to offset: *(Strike Enter to exit this command)*

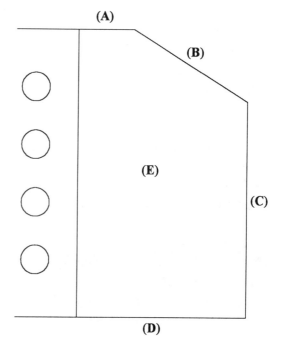

Step #6

Use the Trim command to partially delete the horizontal and vertical segments labeled "A", "B", "C", and "D" at the right. Select the three dashed entities at the right as cutting edges.

Command: **Trim**
Select cutting edge(s)...
Select objects: *(Select the three dashed entities at the right)*
Select objects: *(Strike Enter to continue)*
<Select object to trim>/Undo: *(Select the short vertical line segment at "A")*
<Select object to trim>/Undo: *(Select the horizontal line segment at "B")*
<Select object to trim>/Undo: *(Select the short vertical line segment at "C")*
<Select object to trim>/Undo: *(Select the horizontal line segment at "D")*
<Select object to trim>/Undo: *(Strike Enter to exit this command)*

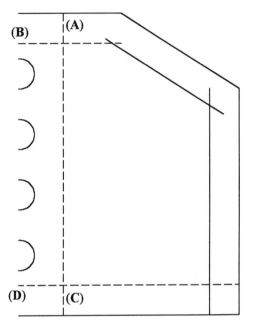

Step #7

Use the Fillet command to create corners at the intersections of the four line segments illustrated at the right. The Fillet radius by default is set to "zero" to accomplish this unless it has been set to a positive value. The Trim command could also be used to accomplish this step.

Command: **Fillet**
Polyline/Radius/<Select first object>: *(Select line "A")*
Select second object: *(Select line "B")*

Command: **Fillet**
Polyline/Radius/<Select first object>: *(Select line "B")*
Select second object: *(Select line "C")*

Command: **Fillet**
Polyline/Radius/<Select first object>: *(Select line "C")*
Select second object: *(Select line "D")*

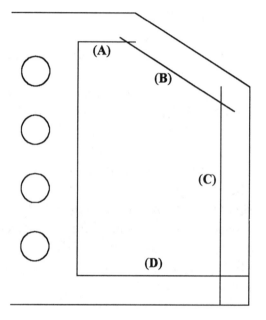

Checking the Accuracy of Fitting.Dwg

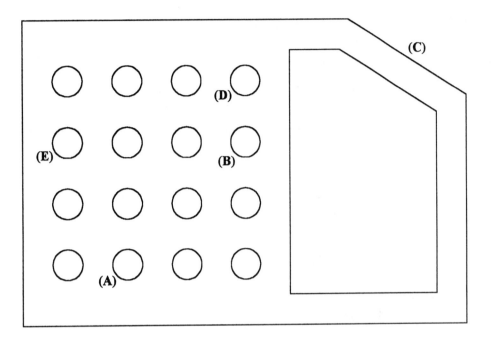

Once the Fitting has been constructed, answer the questions below to determine the accuracy of the drawing. Use the illustration above to assist in answering the questions.

1. The total area of the Fitting with the inner slot and all holes removed is _____

2. The total distance from the center of circle "A" to the center of circle "B" is _____

3. The total length of line "C" is _____

4. The angle formed in the X-Y plane from the center of circle "D" to the center of circle "E" is

_____

5. The absolute coordinate value of the center of circle "D" is _____

A solution for each question follows complete with the method used to arive at the answer. Apply these methods to different objects with similar needs.

Question #1
The total area of the Fitting with all three holes removed is ?

Refer to the illustration below and the text at the right for the correct procedure used in determining the area of the Fitting.

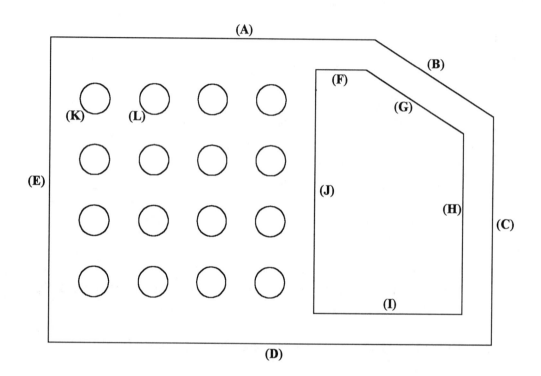

Question #1
The total area of the Fitting with the inner slot and all holes removed is ?

The Area command will be used to first calculate the total area of the object and then subtract the slot and all holes. However before using the Area command, all entities representing the outline of the Fitting and slot must be converted into a polyline. The Pedit command with the Join option is used to best accomplish this. Use the illustration of the Fitting at the right to guide you in the use of the Pedit and Area commands.

Command: **Pedit**
Select polyline: *(Select the entity labeled "A")*
Entity selected is not a polyline.
Do you want to turn it into one? <Y>: *(Strike Enter to continue)*
Close/Join/Width/Edit vertex/Fit/Spline/Decurve/Ltype gen/Undo/eXit <X>: **Join**
Select objects: *(Select lines "B", "C", "D", and "E")*
Select objects: *(Strike Enter to continue)*
4 segments added to polyline
Open/Join/Width/Edit vertex/Fit/Spline/Decurve/Ltype gen/Undo/eXit <X>: *(Strike Enter to exit this command)*

All outer entities now consist as one polyline entity. Repeat the above Pedit procedure for converting entities "F", "G", "H", "I", and "J" into one polyline. Now the Area command may be successfully used to calculate the area of the object with theslot and all holes removed.

Command: **Area**
<First point>/Entity/Add/Subtract: **Add**
<First point>/Entity/Subtract: **Entity**
(ADD mode) Select circle or polyline: *(Select the edge of the Fitting near "A")*
Area = 36.25, Perimeter = 24.11
Total area = 36.25
(ADD mode) Select circle or polyline: *(Strike Enter to continue)*
<First point>/Entity/Subtract: **Subtract**
<First point>/Entity/Add: **Entity**
(SUBTRACT mode) Select circle or polyline: *(Select the slot near "F")*
Area = 9.16,   Perimeter = 12.27
Total area = 27.09
(SUBTRACT mode) Select circle or polyline: *(Select circle"K")*
Area = 0.20,   Circumference = 1.57
Total area = 26.90
(SUBTRACT mode) Select circle or polyline: *(Select circle"L")*
Area = 0.20,   Circumference = 1.57
Total area = 26.70
(SUBTRACT mode) Select circle or polyline: *(Carefully select the remaining 14 circles individually)*
Total area = 23.95
(SUBTRACT mode) Select circle or polyline: *(Strike Enter to continue)*
<First point>/Entity/Add: *(Strike Enter to exit this command)*

The total area of the Fitting with the slot and all holes removed is <u>23.95</u>

Question #2
The total distance from the center of circle "A" to the center of circle "B" is ?

Use the Dist (Distance) command to calculate the distance from the center of circle "A" to the center of circle "B". Be sure to use the Osnap-Center mode. Notice that additional information is given when using the Dist command. For the purpose of this question, we will only be looking for the distance.

Command: **Dist**
First point: **Cen**
of *(Select the edge of circle "A")*
Second point: **Cen**
of *(Select the edge of circle "B")*
Distance = 2.83

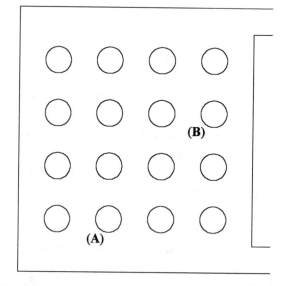

The total distance from the center of circle "A" to the center of circle "B" is 2.83

Question #3
The total length of line "C" is ?

Use the Dist (Distance) command to find the total length of line "C". Be sure to use the Osnap-Endpoint mode. Notice that additional information is given when using the Dist command. For the purpose of this question, we will only be looking for the distance. The List command could also be used to perform this operation. However since the outline of the Fitting consists of one continuous polyline segment, the polylines would have to be converted into individual line segments using the Explode command before using the list command on segment "C".

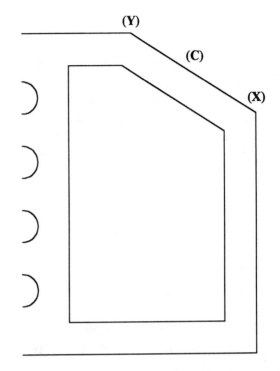

Command: **Dist**
First point: **Endp**
of *(Select the endpoint of the line at "X")*
Second point: **Endp**
of *(Select the endpoint of the line at "Y")*
Distance = 2.36

The total length of line "C" is <u>2.36</u>

Question #4
The angle formed in the X-Y plane from the center of circle "D" to the center of circle "E" is ?

Use the Dist (Distance) command to calculate the angle from the center of circle "D" to the center of circle "E". Be sure to use the Osnap-Center mode. Notice that additional information is given when using the Dist command. For the purpose of this question, we will only be looking for the angle in the X-Y plane.

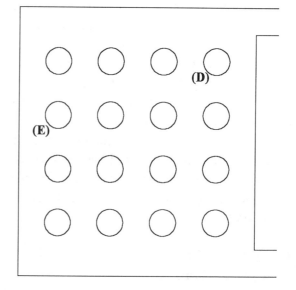

Command: **Dist**
First point: **Cen**
of *(Select the edge of circle "D")*
Second point: **Cen**
of *(Select the edge of circle "E")*
Angle in X-Y Plane = 198

The angle formed in the X-Y plane from the center of circle "D" to the center of circle "E" is <u>198 degrees</u>

Question #5
The absolute coordinate value of the center of circle "D" is ?

The ID command is used to get the current absolute coordinate information on a desired point. This command will display the X, Y, and Z coordinate values. Since this is a 2 dimensional problem, only the X and Y values will be used.

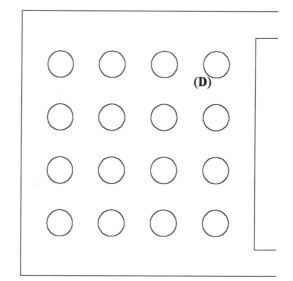

Command: **ID**
Point: **Cen**
of *(Select the edge of circle "D")*
X = 5.99, Y = 5.91, Z = 0.000

The absolute coordinate value of the center of circle "D" is <u>5.99,5.91</u>

Additional Questions for Fitting.Dwg

Suppose the Fitting is rotated at a -10 degree angle using "X" as the center of rotation. Follow the command sequence below to perform this operation.

Command: **Rotate**
Select objects: **All**
Select objects: *(Strike Enter to continue)*
Base point: **Endp**
of *(Select the inclined line near "X")*
<Rotation angle>/Reference: **-10**

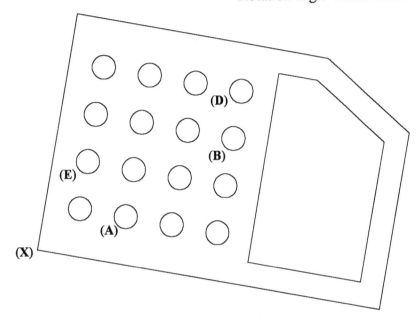

Once the Fitting has been rotated, answer the questions below to determine the accuracy of the drawing. Use the illustration above to assist in answering the questions.

1. The absolute coordinate value of the center of circle "D" is _____

2. The absolute coordinate value of the center of circle "A" is_____

3. The absolute coordinate value of the center of circle "B" is_____

4. The angle formed in the X-Y plane from the center of circle "D" to the center of circle "E" is

_____

A solution for each question follows complete with the method used to arive at the answer. Apply these methods to different objects with similar needs.

Question #1
The absolute coordinate value of the center of circle "D" is ?

The ID command is used to get the current absolute coordinate information on a desired point. This command will display the X, Y, and Z coordinate values. Since this is a 2 dimensional problem, only the X and Y values will be used.

Command: **ID**
Point: **Cen**
of *(Select the edge of circle "D")*
X = 6.63, Y = 5.20, Z = 0.000

The absolute coordinate value of the center of circle "D" is <u>6.63,5.20</u>

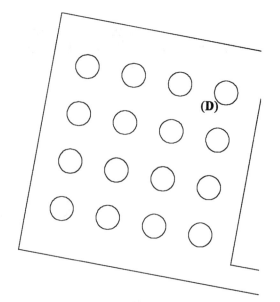

Question #2
The absolute coordinate value of the center of circle "A" is ?

The ID command is used to get the current absolute coordinate information on point "A". This command will display the X, Y, and Z coordinate values. Since this is a 2 dimensional problem, only the X and Y values will be used.

Command: **ID**
Point: **Cen**
of *(Select the edge of circle "A")*
X = 4.14, Y = 2.59, Z = 0.000

The absolute coordinate value of the center of circle "A" is <u>4.14,2.59</u>

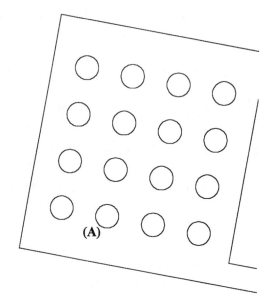

Question #3
The absolute coordinate value of the center of circle "B" is ?

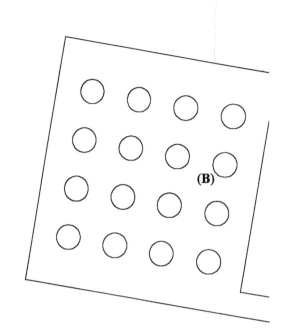

The ID command is used to get the current absolute coordinate information on point "B". This command will display the X, Y, and Z coordinate values. Since this is a 2 dimensional problem, only the X and Y values will be used.

Command: **ID**
Point: **Cen**
of *(Select the edge of circle "B")*
X = 6.45, Y = 4.21, Z = 0.000

The absolute coordinate value of the center of circle "B" is 6.45,4.21

Question #4
The angle formed in the X-Y plane from the center of circle "D" to the center of circle "E" is ?

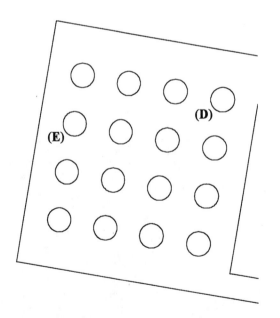

Use the Dist (Distance) command to calculate the angle from the center of circle "D" to the center of circle "E". Be sure to use the Osnap-Center mode. Notice that additional information is given when using the Dist command. For the purpose of this question, we will only be looking for the angle in the X-Y plane.

Command: **Dist**
First point: **Cen**
of *(Select the edge of circle "D")*
Second point: **Cen**
of *(Select the edge of circle "E")*
Angle in X-Y Plane = 188

The angle formed in the X-Y plane from the center of circle "D" to the center of circle "E" is 188 degrees

Notes

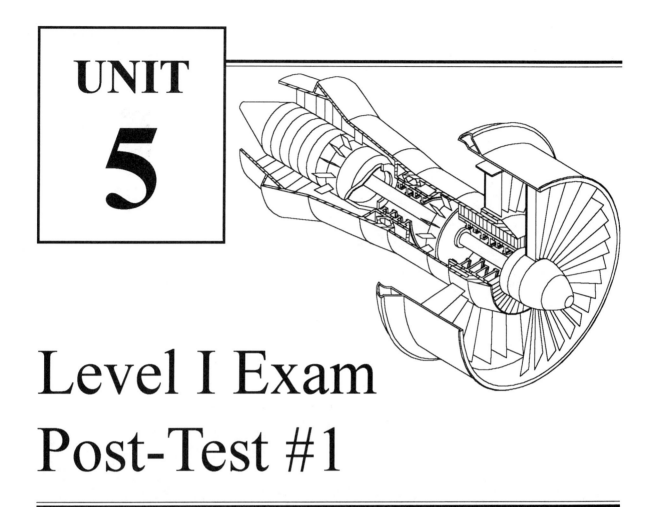

UNIT

5

Level I Exam
Post-Test #1

This first Level I Certification Exam Post-Test has been developed for more practice in preparing for the full 3-hour AutoCAD Level I Certification Exam. Post-Test #1 is designed to be completed in 1 hour and 42 minutes, which represents a little over half of the actual exam. The AutoCAD Level I Certification Exam Post-Test #1 consists of the following two segments:

- The first segment concentrates on drawing skills. Three problems will be drawn and the questions answered in a total of 72 minutes.

- The second segment concentrates on general AutoCAD knowledge in the form of 38 multiple-choice questions. These 38 questions will be answered in 30 minutes.

Work through Post-Test #1 at a good pace paying strict attention to the amount of time spent on each problem and multiple-choice question. Answers for all Post-Test #1 question are located in Unit 7, page 139.

Notes

Level I Exam Post-Test #1 Section I Drawing Segment

Construct all three drawing problems and answer the questions that follow each problem. The problems may be completed in any order. You should allow yourself a total of 72 minutes to complete all three problems.

When all problems have been completed and time still remains, use the extra time to carefully check your answers.

Problem 1

Plate1.Dwg

Directions for Plate1.Dwg

Create a drawing of Plate1 below according to the following instructions:

Use the Limits command and set the upper right corner of the screen area to a value of 36.000,24.000. Use the Units command and set to decimal units. Set the number of digits to the right of the decimal point from 4 to 3. Accept the defaults for the remaining prompts.

Begin this drawing by placing the center of the 4.000 diameter circle at coordinate (16.000,13.000).

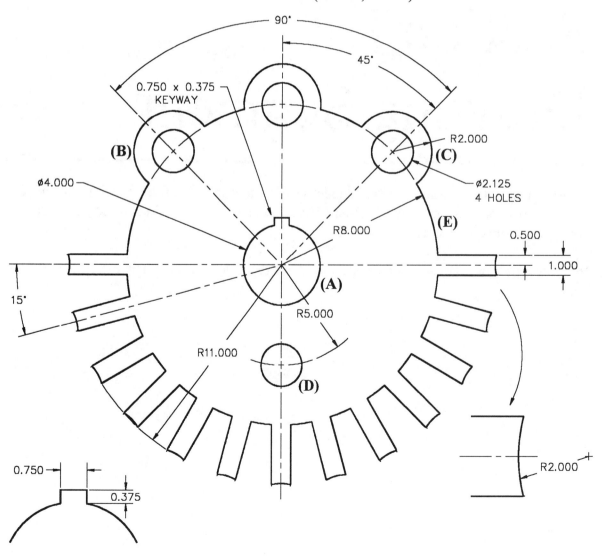

Questions for Plate1.Dwg

Refer to the drawing of Plate1 on the previous page to answer questions #1 through #5 below:

1. The distance from the center of the 2.000 radius arc "B" to the center of the 2.000 radius arc "C" is
   - (A) 10.286
   - (B) 10.293
   - (C) 11.300
   - (D) 11.307
   - (E) 11.314

2. The absolute coordinate value of the center of arc "C" is
   - (A) 21.657,18.657
   - (B) 21.657,18.664
   - (C) 21.657,18.671
   - (D) 21.657,18.678
   - (E) 21.657,18.685

3. The angle formed in the X-Y plane from the center of the 2.000 radius arc "C" to the center of 2.125 diameter hole "D" is
   - (A) 242 degrees
   - (B) 210 degrees
   - (C) 175 degrees
   - (D) 136 degrees
   - (E) 118 degrees

4. The total length of arc "E" is
   - (A) 3.766
   - (B) 3.772
   - (C) 3.778
   - (D) 3.784
   - (E) 3.790

5. The total area of Plate1 with all holes including keyway removed is
   - (A) 232.259
   - (B) 232.265
   - (C) 232.271
   - (D) 232.277
   - (E) 232.283

Provide the answers in the spaces below:

1. _____

2. _____

3. _____

4. _____

5. _____

CONTINUE ON TO THE NEXT PAGE...

Problem 2

Lever2.Dwg

Directions for Lever2.Dwg

Use the Units command to set the units to decimal. Keep the number of digits to the right of the decimal point at 4 places. Keep the remaining default values.

Begin this drawing by locating the center of the 1.0000 diameter circle at coordinate (2.2500,4.0000).

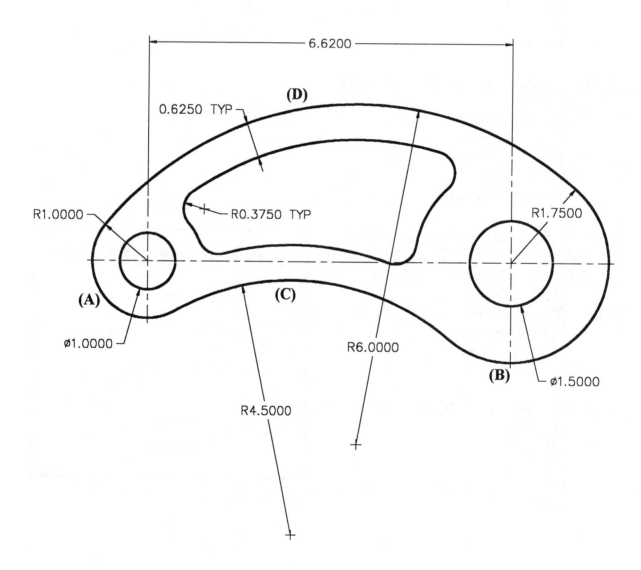

Questions for Lever2.Dwg

Refer to the drawing of Lever2 on the previous page to answer questions #6 through #10 below:

6. The distance from the center of the 1.0000 radius arc "A" to the intersection of the circle and center line at "B" is closest to
    (A) 6.8456
    (B) 6.8462
    (C) 6.8474
    (D) 6.8480
    (E) 6.8486

7. The absolute coordinate value of the center of the 4.5000 radius arc "C" is
    (A) 4.8944,-.08226
    (B) 4.8944,-0.8226
    (C) 4.8950,-0.8232
    (D) 4.8956,-0.8238
    (E) 4.8962,-0.8244

8. The total length of arc "C" is
    (A) 5.3583
    (B) 5.3589
    (C) 5.3595
    (D) 5.3601
    (E) 5.3607

9. The total area of Lever2 with the inner irregular shape and both holes removed is
    (A) 17.6813
    (B) 17.6819
    (C) 17.6825
    (D) 17.6831
    (E) 17.6837

10. The angle formed in the X-Y plane from the quadrant of arc "D" to the center of the 1.5000 circle is
    (A) 315 degrees
    (B) 259 degrees
    (C) 201 degrees
    (D) 135 degrees
    (E) 61 degrees

Provide the answers in the spaces below:

6._____

7._____

8._____

9._____

10._____

CONTINUE ON TO THE NEXT PAGE...

Problem 3

Site1.Dwg

Directions for Site1.Dwg

Use the Units command to set the units to engineering. Set the number of digits to the right of the decimal point from 4 to 2. Begin Site1 with the intersection of "A" located at absolute coordinate (50'-0.00",10'-0.00").

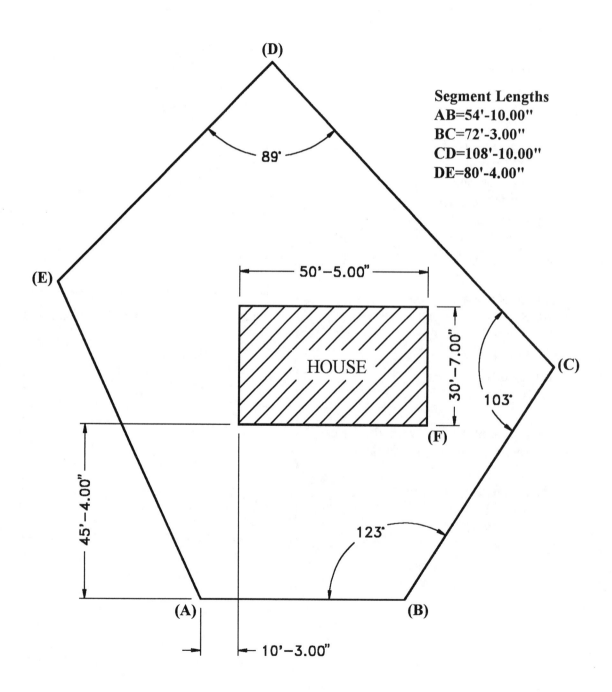

Segment Lengths
AB=54'-10.00"
BC=72'-3.00"
CD=108'-10.00"
DE=80'-4.00"

Questions for Site1.Dwg

Refer to the drawing of Site1 on the previous page to answer questions #11 through #15 below:

11. The total length of line segment "AE" is closest to
    (A) 88'-6.00"
    (B) 89'-0.00"
    (C) 89'-6.00"
    (D) 90'-0.00"
    (E) 90'-6.00"

12. The X,Y coordinate value of vertex "C" is closest to
    (A) 144'-2.00",69'-7.00"
    (B) 144'-2.00",70'-0.00"
    (C) 144'-2.00",70'-7.00"
    (D) 144'-9.00",70'-7.00"
    (E) 145'-2.00",70'-7.00"

13. The angle formed in the X-Y plane from the intersection of the House at "F" to the intersection of vertex "A" is closest to
    (A) 214 degrees.
    (B) 217 degrees.
    (C) 220 degrees.
    (D) 223 degrees.
    (E) 226 degrees.

14. The total area in square feet of the site plan with the house removed is closest to
    (A) 9411 sq ft.
    (B) 9432 sq. ft.
    (C) 9486 sq. ft.
    (D) 9512 sq. ft.
    (E) 9530 sq. ft.

15. A mistake was made in the layout of the original survey of the site plan; vertex "D" must be stretched 20'-0.00" straight down from its present position. The new area in square feet of the site plan with the house removed is closest to
    (A) 8188 sq. ft.
    (B) 8209 sq. ft.
    (C) 8235 sq. ft.
    (D) 8251 sq. ft.
    (E) 8279 sq. ft.

Provide the answers in the spaces below:

11. _____

12. _____

13. _____

14. _____

15. _____

END OF SECTION I - DO NOT PROCEED FURTHER UNTIL TOLD TO DO SO

Notes

Level I Exam Post-Test #1 Section II General Knowledge Segment

Answer each of the 38 multiple-choice questions. The questions may be answered in any order. It is considered good practice to answer the easier questions first. If a question seems difficult, do not waste time trying to answer it. Go on to the next question and come back to the difficult question or questions later. Be sure to provide the best possible answer for each question.

You should allow yourself a total of 30 minutes to answer all 38 multiple-choice questions.

Place the best answer in the appropriate box for each of the following questions. Unless otherwise specified, all questions are "Single Answer Multiple Choice".

16. The command used to automatically repeat a command is
    (A) NEXT.
    (B) REPEAT.
    (C) MULTIPLE.
    (D) COMREPT.
    (E) REDO.

17. By default, pressing the "Enter" button of a mouse or digitizing puck while holding down the Shift key activates
    (A) POP 0
    (B) a pull-down menu containing all Object Snap modes.
    (C) the Enter key.
    (D) Only A and B.
    (E) Only B and C.

18. The dimensioning subcommand that lists all dimension variables with their current values is called
    (A) DIMVAR.
    (B) STATUS.
    (C) VARIABLE.
    (D) DIMSTAT.
    (E) None of the above.

19. Redefining a block performs an automatic REGEN of the display screen when
    (A) REGENAUTO is turned On.
    (B) REGENAUTO is turned Off.
    (C) REGEN is turned On.
    (D) Both B and C.
    (E) None of the above.

|← 2.07 →|← 1.93 →|← 2.00 →|

20. The dimensioning subcommand used to dimension the object illustrated above is
    (A) BASELINE.
    (B) CONTINUOUS.
    (C) RADIAL.
    (D) LEADER.
    (E) ORDINATE.

Question 21 is considered a "Short Answer or Fill In the Blank" question. Provide the correct answer in the space provided that satisfies the particular question.

21. The command used to view a different portion of a drawing without changing its magnification is

    _____

22. To draw a line at a distance of sixteen feet, four and five eights inches in the direction directly above the last known point, enter
    (A) @-16'4-5/8<90
    (B) @16'-4-5/8<90
    (C) @16'-4 5/8<90
    (D) 16'-4-5/8<90
    (E) @16'4-5/8<90

23. The command used to identify a new insertion point for a block or drawing file is
    (A) BASE.
    (B) INSERT.
    (C) NEW.
    (D) POINT.
    (E) OSNAP-Insert option.

24. A block may not be exploded if
    (A) it was inserted into a drawing with negative absolute coordinates.
    (B) it was inserted into a drawing with a negative rotation value.
    (C) it was inserted into a drawing with positive absolute coordinates.
    (D) it was inserted into a drawing with different X and Y scale factor values.
    (E) Both A and D

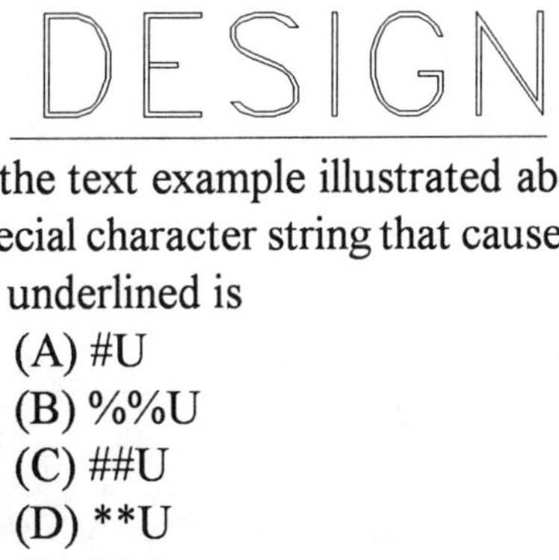

DESIGN

25. In the text example illustrated above, the special character string that causes text to be underlined is
    (A) #U
    (B) %%U
    (C) ##U
    (D) **U
    (E) %U

26. The text option that prompts the user for two endpoints and then automatically computes the text height so the text is positioned between the two points is
    (A) Centered.
    (B) Fully Centered.
    (C) Right Justified.
    (D) Aligned.
    (E) Fit.

27. The "C" option of the LINE command stands for
    (A) Close.
    (B) CENter.
    (C) Create.
    (D) Cling.
    (E) None of the above.

Question 28 is considered a "Multiple-Answer Multiple-Choice question. Supply all possible answers that satisfy the particular question.

28. Valid hatching styles include
    (A) Normal
    (B) "O"
    (C) "I"
    (D) "U"
    (E) "S"

29. When using the ARC command with the "S,C,A" option, the letter "C" stands for
    (A) circle.
    (B) circumference.
    (C) chord.
    (D) center.
    (E) None of the above.

30. The POLYGON command prompts the user and constructs the polygon relative to a circle's
    (A) quadrant.
    (B) diameter.
    (C) radius.
    (D) area.
    (E) perimeter.

CONTINUE ON TO THE NEXT PAGE...

Question 31 is considered a "Multiple-Answer Multiple-Choice question. Supply all possible answers that satisfy the particular question.

31. The following entities affected by the FILL command include
    - (A) donuts.
    - (B) polylines.
    - (C) areas.
    - (D) circles.
    - (E) dimension arrowheads.

32. Valid object selection set modes are
    - (A) Window and Extents.
    - (B) Extents and All.
    - (C) Limits and Box.
    - (D) Last and Limits.
    - (E) Fence and Window Polygon.

33. For selecting the object to trim, the following selection set options may be used:
    - (A) Window.
    - (B) Crossing.
    - (C) Box.
    - (D) Fence.
    - (E) All of the above.

Question 34 is considered a "Short Answer or Fill In the Blank" question. Provide the correct answer in the space provided that satisfies the particular question.

34. The command that allows you to make multiple copies of existing objects in a rectangular or circular pattern is

_____

Questions 35 and 36 are considered "Multiple-Answer Multiple-Choice questions. Supply all possible answers that satisfy the particular question.

35. The following entities that may be used as valid boundary edges when using the EXTEND command include
    - (A) Blocks.
    - (B) Circles.
    - (C) Polylines.
    - (D) unexploded Hatch patterns.
    - (E) Arcs.

36. Entities that can be successfully exploded include
    - (A) Blocks.
    - (B) Associative Dimensions.
    - (C) Polylines.
    - (D) Minserted Blocks.
    - (E) Blocks inserted at different scale factors.

37. The CHANGE command has no effect on
    - (A) lines.
    - (B) blocks.
    - (C) layers.
    - (D) views.
    - (E) linetypes.

38. To reproduce an existing entity 5 times in a full circular pattern, use the polar option of the ARRAY command with the number of items and angle of
    - (A) 5 and 180 respectively.
    - (B) 5 and 360 respectively.
    - (C) 6 and 90 respectively.
    - (D) 6 and 180 respectively.
    - (E) 6 and 360 respectively.

Question 39 is considered a "Multiple-Answer Multiple-Choice question. Supply all possible answers that satisfy the particular question.

39. Of the following entities, those that may be successfully modified using the STRETCH command include
    (A) text.
    (B) lines.
    (C) solids.
    (D) polylines.
    (E) unexploded hatch patterns.

40. The DIVIDE command
    (A) divides an entity into equal parts.
    (B) places markers along the entity at the divide points.
    (C) uses the current point style set by PDMODE.
    (D) uses the current point size set by PDSIZE.
    (E) All of the above are true of the DIVIDE command.

Question 41 is considered a "Multiple-Answer Multiple-Choice question. Supply all possible answers that satisfy the particular question.

41. Valid grip modes include
    (A) Copy.
    (B) Trim.
    (C) Rotate.
    (D) Break.
    (E) Extend.

42. The command that lets you access information about an entity is
    (A) DDMODIFY.
    (B) DDINQUIRE.
    (C) LIST.
    (D) Both A and B.
    (E) Both A and C.

43. The command used to obtain the absolute coordinate value of a specific point is
    (A) INQUIRE.
    (B) FIND.
    (C) ANALYSIS.
    (D) STATUS.
    (E) ID.

44. A layer that is turned off will be
    (A) seen but not plotted.
    (B) plotted but not seen.
    (C) the current layer.
    (D) neither seen nor plotted.
    (E) Both plotted and seen.

Question 45 is considered a "Multiple-Answer Multiple-Choice question. Supply all possible answers that satisfy the particular question.

45. Valid options of the DDLMODES dialogue box include
    (A) Load.
    (B) Lock.
    (C) ON.
    (D) Freeze.
    (E) Color.

CONTINUE ON TO THE NEXT PAGE...

46. To set up multiple viewport windows, use the
    (A) VPORTS command.
    (B) VIEWPORT command.
    (C) VIEWPORTS command.
    (D) Only A and C.
    (E) Only A and B.

47. The OSNAP "Quadrant" mode is used for
    (A) finding the quadrant of a rectangle or polygon.
    (B) snapping to the closest quadrant point of a circle.
    (C) snapping to a distant point.
    (D) finding certain cartesian coordinates.
    (E) All of the above.

48. The command used to adjust the size of the object snap target box is
    (A) TARGET.
    (B) APERTURE.
    (C) DDOSNAP (the slider bar area).
    (D) Both A and C.
    (E) Both B and C.

Question 49 is considered a "Short Answer or Fill In the Blank" question. Provide the correct answer in the space provided that satisfies the particular question.

49. The object snap mode used to assist in the construction of the line segments in the illustration above is

Questions 50, 51, and 52 are considered "Multiple-Answer Multiple-Choice questions. Supply all possible answers that satisfy the particular question.

50. The following items that can be changed when setting plot parameters include
    (A) hidden line removal.
    (B) pen numbers.
    (C) the linetype scale.
    (D) plot origin.
    (E) plotting to a file.

51. The following items that can be successfully renamed using the DDRENAME dialogue box include
    (A) Wblock file names.
    (B) External reference file names.
    (C) Block names.
    (D) Text style names.
    (E) User coordinate system names.

52. Valid options of the FILES command include
    (A) deleting files.
    (B) unlocking files.
    (C) editing files.
    (D) renaming files.
    (E) listing files.

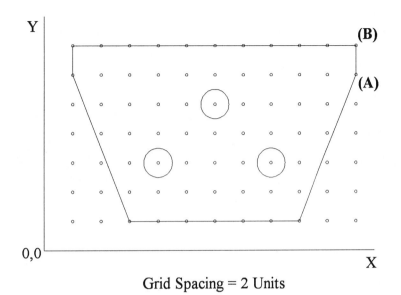

Grid Spacing = 2 Units

53. In the figure above, the polar coordinates
of Point "A" from Point "B" are
(A) @2.00<360
(B) @2.00<270
(C) @2.00<180
(D) @2.00<90
(E) @2.00<0

END OF SECTION II

Notes

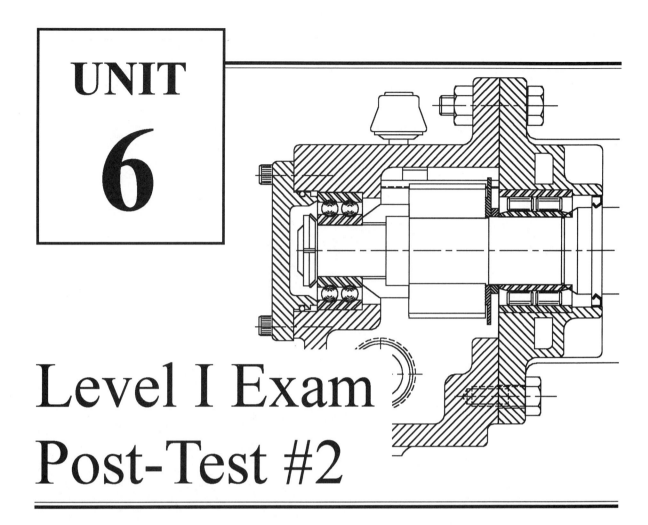

UNIT

6

Level I Exam
Post-Test #2

Use this second Level I Certification Exam Post-Test for more practice in preparing for the full 3-hour AutoCAD Level I Certification Exam. Post-Test #2 is designed to be completed in a time frame similar to the first post test; namely in 1 hour and 42 minutes, which represents a little over half of the actual exam. The AutoCAD Level I Certification Exam Post-Test #2 consists of the following two segments:

- The first segment concentrates on drawing skills. Three problems will be drawn and the questions answered in a total of 72 minutes.

- The second segment concentrates on general AutoCAD knowledge in the form of 38 multiple-choice questions. These 38 questions will be answered in 30 minutes.

Work through Post-Test #2 at a good pace paying strict attention to the amount of time spent on each problem and multiple-choice question. Answers for all Post-Test #2 question are located in Unit 7, page 140.

Notes

Level I Exam Post-Test #2 Section I Drawing Segment

Construct all three drawing problems and answer the questions that follow each problem. The problems may be completed in any order. You should allow yourself a total of 72 minutes to complete all three problems.

When all problems have been completed and time still remains, use the extra time to carefully check your answers.

Problem 1

Plan1.Dwg

Directions for Plan1.Dwg

Start a new drawing called Plan1. Change from decimal units to architectural units. Keep all remaining default values. **All wall thicknesses measure 4"**. Answer the questions on the next page regarding this drawing.

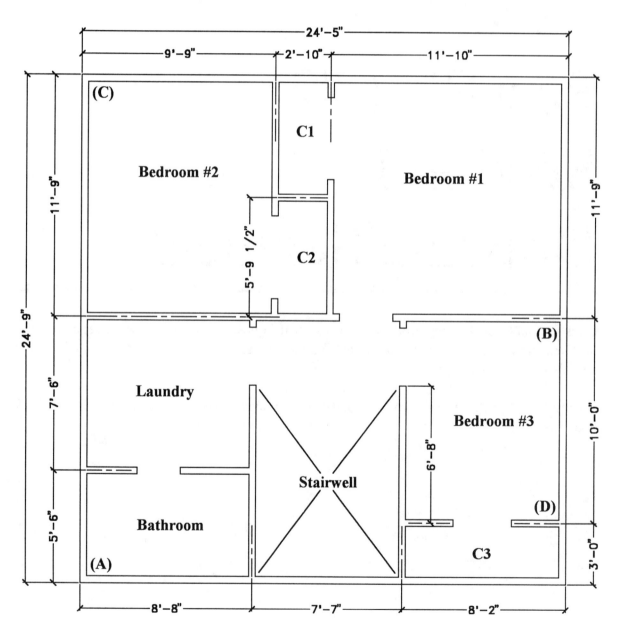

SECOND FLOOR

Questions for Plan1.Dwg

Refer to the drawing of Plan1 on the previous page to answer questions #1 through #5 below:

1. The total area in square feet of all bedrooms is closest to
      (A) 296 sq. ft.
      (B) 306 sq. ft.
      (C) 316 sq. ft.
      (D) 326 sq. ft.
      (E) 336 sq. ft.

2. The distance from the intersection of the corner at "A" to the intersection of the corner at "B" is closest to
      (A) 25'-2"
      (B) 25'-7"
      (C) 26'-0"
      (D) 26'-5"
      (E) 26'-10"

3. The total area in square feet of all closets (C1, C2, and C3) is closest to
      (A) 46 sq. ft.
      (B) 48 sq. ft.
      (C) 50 sq. ft.
      (D) 52 sq. ft.
      (E) 54 sq. ft.

4. The total area in square feet of the laundry and bathroom is closest to
      (A) 91 sq. ft.
      (B) 93 sq. ft.
      (C) 95 sq. ft.
      (D) 97 sq. ft.
      (E) 99 sq. ft.

5. The distance from the intersection of the corner at "C" to the intersection of the corner at "D" is closest to
      (A) 31'-0"
      (B) 31'-5"
      (C) 31'-10"
      (D) 32'-3"
      (E) 32'-8"

Provide the answers in the spaces below:

1. _____

2. _____

3. _____

4. _____

5. _____

CONTINUE ON TO THE NEXT PAGE...

Problem 2

Pattern5.Dwg

Directions for Pattern5.Dwg

Begin the construction of Pattern5 illustrated below by keeping the default drawing limits set to 0,0 for the lower left corner and 250,200 for the upper right corner. Keep the default units set to decimal but change the number of decimal places past the zero from 4 to 0.

Place Vertex "A" at absolute coordinate (190,30).

Line AB is a straight orthogonal segment.

Segment Lengths
AB=94
BC=40
CD=35
DE=57
EF=82
FG=61
GH=73
HJ=43

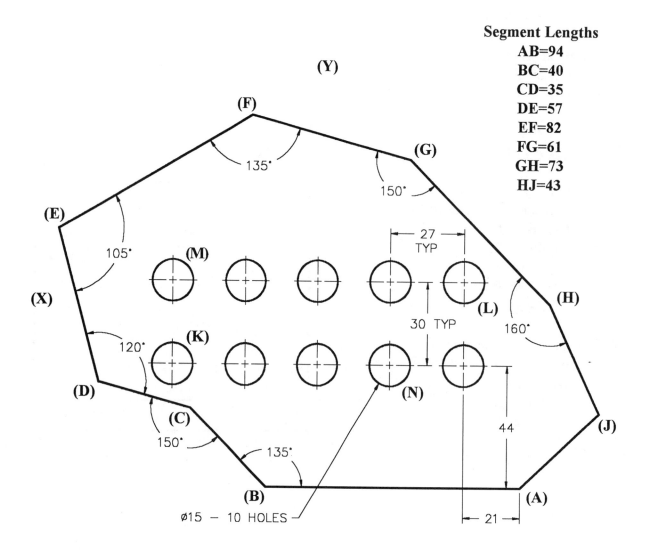

Questions for Pattern5.Dwg

Refer to the drawing of Pattern5 on the previous page to answer questions #6 through #10 below:

6. The distance from the Point "J" to Point "A" is
   - (A) 38
   - (B) 39
   - (C) 40
   - (D) 41
   - (E) 42

7. The total area of Pattern5 with all 10 holes removed is
   - (A) 16369
   - (B) 16370
   - (C) 16371
   - (D) 16372
   - (E) 16373

8. The distance from the center of the 15 diameter hole "K" to the center of the 15 diameter hole "L" is
   - (A) 109
   - (B) 110
   - (C) 111
   - (D) 112
   - (E) 113

9. The angle formed in the X-Y plane from the center of the 15 diameter hole "M" to the center of the 15 diameter hole "N" is closest to
   - (A) 340 degrees.
   - (B) 343 degrees.
   - (C) 246 degrees.
   - (D) 349 degrees.
   - (E) 352 degrees.

10. Use the Stretch command to lengthen Pattern5. Use "X" as the first point of the stretch crossing box. Use "Y" as the other corner. Pick the intersection at "F" as the base point and stretch Pattern5 a total of 23 units in the 180 direction. The new total area of Pattern5 with all 10 holes removed is
    - (A) 17746
    - (B) 17753
    - (C) 17760
    - (D) 17767
    - (E) 17774

Provide the answers in the spaces below:

6. _____

7. _____

8. _____

9. _____

10. _____

CONTINUE ON TO THE NEXT PAGE...

Problem 3

Coupler.Dwg

Directions for Coupler.Dwg

Use the Units command to change the number of decimal places past the zero from 4 to 2. Keep the remaining Units command default values.

Begin by drawing the center of the 150.00 mm diameter circle at absolute coordinate (180.00,120.00).

The four sides of the rectangular slots are composed of two arcs and two parallel lines. The two parallel lines are spaced 5.00 mm apart from each other.

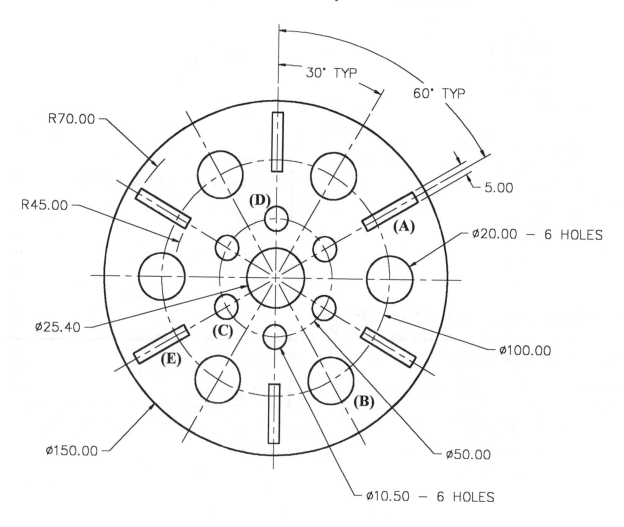

Questions for Coupler.Dwg

Refer to the drawing of the Coupler on the previous page to answer questions #11 through #15 below:

11. Move the entire drawing a polar distance of 15.00 mm at an angle of 5 degrees. The new X, Y coordinate value of the center of the coupler is
    (A) 197.56,123.48
    (B) 194.94,121.31
    (C) 184.94,122.38
    (D) 176.91,112.38
    (E) 169.72,101.27

12. The center-to-center distance between any two adjacent 20.00 mm diameter holes is
    (A) 55.00
    (B) 50.00
    (C) 45.00
    (D) 40.00
    (E) 35.00

13. The surface area of the coupler with all holes and slots removed is closest to
    (A) 14,737.00 sq mm
    (B) 14,516.00 sq mm
    (C) 14,010.00 sq mm
    (D) 13,667.00 sq mm
    (E) 13,326.00 sq mm

14. The angle formed in the X-Y plane from the midpoint of the line "A" to the center of 20 mm diameter hole "B" is
    (A) 238 degrees.
    (B) 241 degrees.
    (C) 244 degrees.
    (D) 247 degrees.
    (E) 250 degrees

15. Change all 20.00 mm diameter holes to 25.00 mm diameter holes. The new surface area of the coupler with all holes and slots removed is closest to
    (A) 12,911.00 sq mm
    (B) 12,936.00 sq mm
    (C) 12,950.00 sq mm
    (D) 12,991.00 sq mm
    (E) 13,033.00 sq mm

Provide the answers in the spaces below:

11._____

12._____

13._____

14._____

15._____

END OF SECTION I - DO NOT PROCEED FURTHER UNTIL TOLD TO DO SO

Notes

Level I Exam Post-Test #2 Section II General Knowledge Segment

Answer each of the 38 multiple-choice questions. The questions may be answered in any order. It is considered good practice to answer the easier questions first. If a question seems difficult, do not waste time trying to answer it. Go on to the next question and come back to the difficult question or questions later. Be sure to provide the best possible answer for each question.

You should allow yourself a total of 30 minutes to answer all 38 multiple-choice questions.

Place the best answer in the appropriate box for each of the following questions. Unless otherwise specified, all questions are "Single Answer Multiple Choice".

Question 16 is considered a "Multiple-Answer Multiple-Choice question. Supply all possible answers that satisfy the particular question.

16. The following commands that can be used transparently include
    (A) BLOCK.
    (B) ZOOM.
    (C) SNAP.
    (D) LINE.
    (E) UNITS.

17. Prototype drawing selection is found in the
    (A) FILES dialogue box.
    (B) NEW dialogue box.
    (C) OPEN dialogue box.
    (D) APPLOAD dialogue box.
    (C) SAVE dialogue box.

18. The object selection mode that allows the user to select all entities that lie within the boundaries of a polygon is
    (A) W
    (B) F
    (C) CP
    (D) WP
    (E) C

19. Entities mistakenly added to a selection set may be removed from the selection set using
    (A) Replace
    (B) Reinit
    (C) Recover
    (D) Remove
    (E) Entities once made into a selection set may not be removed.

Question 20 is considered a "Multiple-Answer Multiple-Choice question. Supply all possible answers that satisfy the particular question.

20. The following dimension subcommands that produce associative dimensions if the variable DIMASO is On include
    (A) DIAMETER.
    (B) LEADER.
    (C) HORIZONTAL.
    (D) RADIUS.
    (E) VERTICAL.

21. The dimension subcommand used to change the justification and/or rotate the text of an associative dimension is
    (A) DTEXT.
    (B) TEDIT.
    (C) DIMROTATE.
    (D) ROTATE.
    (E) DIMSLANT.

22. The text justification option that prompts the user for two endpoints and the text height so the text is positioned between the two points is
    (A) Centered.
    (B) Fully Centered.
    (C) Right Justified.
    (D) Aligned.
    (E) Fit.

135°

23. In the illustration above, the special character string that causes text to take on the degree symbol is
   (A) #D
   (B) %D
   (C) ##D
   (D) **D
   (E) %%D

24. The option of the DTEXT command used for determining a specific text alignment mode is
   (A) starting point.
   (B) align.
   (C) justify.
   (D) locate.
   (E) None of the above.

25. Text can be edited using the DDEDIT dialogue box while in
   (A) Model Space.
   (B) Paper Space.
   (C) Text Space.
   (D) Both A and B.
   (E) None of the above.

26. The Draw command that prompts the user for "First corner:" and "Other corner:" is
   (A) PLINE
   (B) ZOOM-Window
   (C) RECTANG
   (D) STRETCH-Crossing
   (E) Both B and D

27. The letter "T" in the "TTR" option of the CIRCLE command means
   (A) transparent.
   (B) trace.
   (C) tangent.
   (D) trim.
   (E) tolerance.

28. Acceptable entities that can be used as boundaries for the BHATCH command include
   (A) lines.
   (B) blocks.
   (C) ellipses.
   (D) polygons.
   (E) All of the above.

Question 29 is considered a "Short Answer or Fill In the Blank" question. Provide the correct answer in the space provided that satisfies the particular question.

29. The object selection mode required for best results by the STRETCH command is

   _____

30. The command used to modify the layer of an entity is
   (A) CHANGE.
   (B) ENTITY.
   (C) CHPROP.
   (D) Both A and C.
   (E) Both B and C.

CONTINUE ON TO THE NEXT PAGE...

31. The OOPS command will restore
    (A) entities that were previously placed on frozen layers.
    (B) entities that were removed while frozen.
    (C) entities that disappeared after creating a block.
    (D) entities that were chained to a different layer.
    (E) None of the above.

32. When using the CHAMFER command, the two legs selected to perform the chamfer **cannot** be
    (A) of the same length.
    (B) perpendicular.
    (C) on different layers.
    (D) constructed with different linetypes.
    (E) parallel.

33. The command used to edit text placed on a drawing is
    (A) EDIT-TEXT.
    (B) DDEDIT .
    (C) CHANGE.
    (D) TEXTEDIT.
    (E) Both B and C.

34. If two non-parallel lines are edited using the FILLET commad and a radius of 0 (zero),
    (A) the lines are extended but do not intersect.
    (B) the lines extend to intersect.
    (C) the first line selected extends while the second line selected remains unchanged.
    (D) the first line selected remains unchanged while the second line selected extends.
    (E) both lines remain unchanged.

35. When using the BREAK command on a circle, the break will occur in the
    (A) clockwise direction.
    (B) counterclockwise direction.
    (C) horizontal direction.
    (D) vertical direction.
    (E) angled direction.

36. In the illustration above, the EXTEND command is used to extend all lines to the boundary edge identified by the hidden line. When picking the entities to extend, the **best** selection mode used to accomplish this in the quickest manner is
    (A) by Crossing.
    (B) by Window.
    (C) by Last.
    (D) by Fence.
    (E) to select the lines to extend individually.

Question 37 is considered a "Multiple-Answer Multiple-Choice question. Supply all possible answers that satisfy the particular question.

37. Valid grip settings from the DDGRIPS dialogue box include
    (A) Grip Modes (Stretch, Move, Rotate, Scale, Mirror)
    (B) Grips Enabled/Disabled.
    (C) Grip Colors.
    (D) Enabling Grips within Blocks.
    (E) Grip Size.

38. When plotting different line weights, pen numbers are assigned to
    (A) text styles.
    (B) named views.
    (C) entity linetypes.
    (D) entity layers.
    (E) entity colors.

39. The quickest way to get back to the last zoomed display is by using the ZOOM command and the
    (A) Previous option.
    (B) Last option.
    (C) Window option.
    (D) Center option.
    (E) Extents option.

40. The Window option of the VIEW command allows the user to
    (A) save a view by means of a window.
    (B) select a view by means of a window.
    (C) call up a view listing within a window.
    (D) change to a new viewing point.
    (E) control the number of views defined in a drawing.

41. To insert a block named "PART1" and have the block automatically converted into individual entities,
    (A) enter "/PART1" for the block name.
    (B) enter "?PART1" for the block name.
    (C) place a check in the "Explode" box located in the DDINSERT dialogue box.
    (D) enter "PART1" for the block name.
    (E) enter "(PART1)" for the block name.

42. The property of a block affected by the CHANGE command is
    (A) the name of the block.
    (B) insertion point.
    (C) X scale factor.
    (D) Y scale factor.
    (E) All of the above are affected.

43. The command that can be used to verify the name of the current drawing is
    (A) ?
    (B) LIST
    (C) INQUIRY
    (D) STATUS
    (E) QUERY

44. The ID command lists
    (A) the relative distance from the last known point.
    (B) the type of entity selected.
    (C) the absolute X, Y, and Z coordinate value of a point.
    (D) the layer name for the selected entity
    (E) the angle in the XY plane from the last known point.

Question 45 is considered a "Multiple-Answer Multiple-Choice question. Supply all possible answers that satisfy the particular question.

45. The operations that may be performed when using the DDLMODES dialogue box include
    (A) freezing layers.
    (B) changing entities to another layer.
    (C) creating a new layer.
    (D) assignment of plotter pens to a layer.
    (E) locking a layer.

CONTINUE ON TO THE NEXT PAGE...

```
Layer Name              State      Color    Linetype
0                       On . .     white    CONTINUOUS
A-DIM-5                 On F L     9        CONTINUOUS
A-SYM-2                 On F .     blue     CONTINUOUS
A-SYM-5                 On F .     9        CONTINUOUS
A-TX-1                  . . .      red      CONTINUOUS
A-TX-5                  . F .      9        CONTINUOUS
C-ROOF-4                On F L     green    CONTINUOUS
```

46. In the illustration above of the layer dialogue box, layers "A-DIM-5" and
"C-ROOF-4" are considered
    (A) Locked, Thawed, and On.
    (B) Locked, Frozen, and On.
    (C) Unlocked, Thawed, and Off.
    (D) Unlocked, Frozen, and On.
    (E) Unlocked, Thawed, and On.

47. The Osnap Mode used to snap to the 0, 90,
180, or 270 degree position of a circle or
arc is
    (A) ENDpoint.
    (B) TANgent.
    (C) NEArest.
    (D) CENter.
    (E) QUAdrant.

48. The command that forces lines to be drawn
parallel to the horizontal or vertical axis is
    (A) NORMAL
    (B) PARALLEL
    (C) SNAP
    (D) ORTHO
    (E) VERT/HOR

49. The command used to change the distance
and spacing of a hidden line is
    (A) CHANGE
    (B) LTSCALE
    (C) SCALE
    (D) SPACE
    (E) DASHDIST

50. The command used to create a drawing
interchange file is
    (A) CRINT.
    (B) CREATE.
    (C) DXFOUT.
    (D) EXCHANGE.
    (E) SHARE.

**Question 51 is considered a "Multiple-An-
swer Multiple-Choice question. Supply all
possible answers that satisfy the particular
question.**

51. Items that may be purged from a drawing
file include
    (A) user defined coordinate systems.
    (B) text styles.
    (C) layers.
    (D) views.
    (E) linetypes.

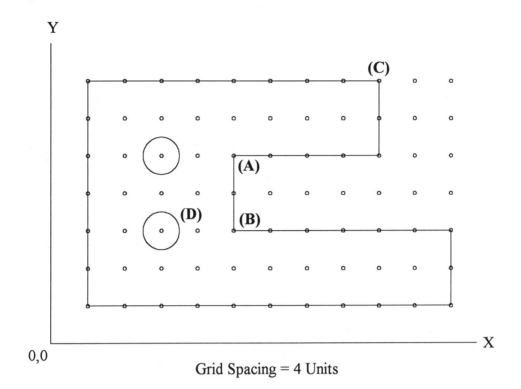

Grid Spacing = 4 Units

52. In the figure above, the polar coordinates of Point "A" from Point "B" are

    (A) @8.00<360
    (B) @8.00<0
    (C) @8.00<90
    (D) @8.00<180
    (E) @8.00<270

53. In the figure above, the relative coordinates of Point "C" from the center of circle "D" are

    (A) @-16.00.-24.00
    (B) @-24.00,-16.00
    (C) @16.00,24.00
    (D) @24.00,16.00
    (E) 16.00,24.00

END OF SECTION II

Notes

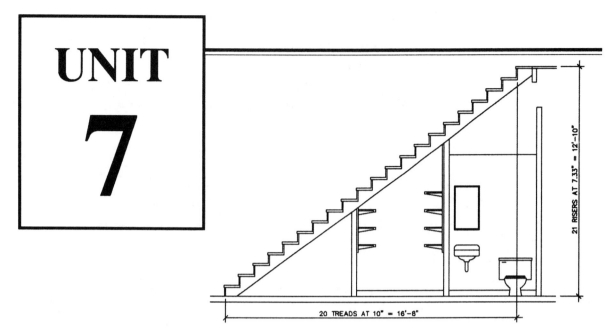

21 RISERS AT 7.33" = 12'-10"

20 TREADS AT 10" = 16'-8"

Level I Exam
Pre/Post-Test Answers

Use this unit to check your success after taking the Level I Pre-Test, Post-Test #1, and Post-Test #2. Once finished with all tests, time yourself again on the same tests and see if you either have completed the problems faster or have correctly answered difficult general knowledge questions.

Answers to the Level I Pre-Test
Drawing Segment
and
General Knowledge Segment

1. A
2. D
3. D
4. D
5. B
6. C
7. E
8. A
9. D
10. B

11. C
12. E
13. E
14. C
15. Dimscale
16. B
17. B
18. D
19. A
20. D
21. A, B, C, E
22. B
23. D
24. A
25. C

Answers to the Level I Post-Test #1
Drawing Segment
and
General Knowledge Segment

1. E
2. A
3. A
4. C
5. B
6. C
7. B
8. C
9. A
10. A
11. E
12. C
13. B
14. D
15. A

16. C
17. D
18. B
19. A
20. B
21. Pan
22. E
23. A
24. D
25. B
26. D
27. A
28. A, B, C, D
29. D
30. C
31. A, B, E
32. E
33. D
34. Array

35. B, C, E
36. A, B, C
37. D
38. E
39. B, C, D
40. E
41. A, C
42. E
43. E
44. D
45. B, C, D, E
46. D
47. B
48. E
49. Tangent
50. A, B, D, E
51. C, D, E
52. A, B, D, E
53. B

Answers to the Level I Post-Test #2
Drawing Segment
and
General Knowledge Segment

1. B
2. E
3. A
4. E
5. C
6. C
7. E
8. D
9. A
10. D
11. B
12. B
13. C
14. E
15. C

16. B, C, E
17. B
18. D
19. D
20. A, C, D, E
21. B
22. E
23. E
24. C
25. D
26. C
27. C
28. E
29. Crossing
30. D
31. C
32. E
33. E
34. B

35. B
36. D
37. B, C, D, E
38. E
39. A
40. A
41. C
42. B
43. D
44. C
45. A, C, E
46. B
47. E
48. D
49. B
50. C
51. B, C, E
52. C
53. D

The
AutoCAD
Certification
Exams

Level II

Notes

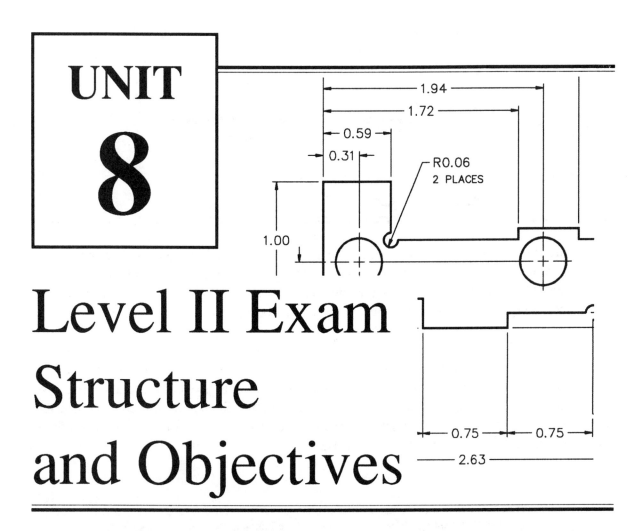

UNIT 8

Level II Exam Structure and Objectives

The AutoCAD Level II Certification Exam is divided into two parts similar to the Level I Certification Exam; a drawing segment and a general knowledge segment. For the drawing segment, the individual will be presented with a test booklet consisting of 4 drawing problems to be completed by the individual. Future drawing segments will consist of a booklet containing 8 drawing problems. Each drawing problem is followed by 5 questions on each problem. The same Inquiry commands used in the Level I Exam are used to analyze each drawing in the Level II Exam; this takes the form of using such commands as Area, Dist, ID, List, and DDMODIFY. An individual should be comfortable with the construction techniques and analysis functions associated with the Region Modeler. The general knowledge segment of the Level II Exam consists of 50 questions followed by 5 possible answers for each question. The individual is to select the best answer for each question.

This unit outlines the structure of the Level II Exam complete with the topics to be tested on, the number of questions per topic, and a topic percentage as it relates to the entire exam. Also, each topic is further outlined with a detailed listing of the objectives an individual must master for successfully passing the Level II Exam. As with the Level I Exam, use these objectives as a study guide for determining strengths and weaknesses and what topics to concentrate on.

Level II Exam Structure Drawing Segment

The AutoCAD Level II Certification Exam is an intense 3 hour examination based on the current release of AutoCAD. Of the 3 total hours spent on the exam, 2 hours are set aside for a drawing segment and 1 hour is set aside for a general AutoCAD knowledge segment. Both exam segments will be scored separately using a statistical analysis method.

The 2 hour drawing segment consists of a test booklet containing 4 drawing problems with 5 multiple choice questions for each problem. Future drawing segments will consist of a booklet containing 8 drawing problems; An individual will select 4 drawings out of the total 8 presented. All drawings in the Level II Exam are designed to be more difficult than the drawings constructed in the Level I Exam. Use the chart below for a breakdown on the categories of questions to be asked and the weight they carry in the Level II Exam.

Level II Exam Drawing Segment Categories	Percentage of This Segment	Number of Questions
Drawing Problem #1	25%	5
Drawing Problem #2	25%	5
Drawing Problem #3	25%	5
Drawing Problem #4	25%	5
Total	100%	20

Level II Exam Structure General Knowledge Segment

The 1 hour Level II general knowledge segment consists of 50 multiple choice questions. The format of the questions will consist of single answer multiple choice questions, multiple answer multiple choice questions, and short answer (fill in the blank) questions. Use the chart below and on the next page for a breakdown on the categories of questions to be asked and the weight they carry in the Level II Exam.

Level II Exam General Knowledge Segment Categories	Percentage of This Segment	Number of Questions
Productivity Techniques	12%	6
User Coordinate Systems	6%	3
External References	6%	3
Model Space/Paper Space	10%	5
Advanced Grips	4%	2
Advanced Selection Sets	4%	2
Region Modeling	4%	2
System Variables	6%	3
Dimension Styles	6%	3
Object Filtering	2%	1
Layer Filtering	2%	1
Wildcards	4%	2

Level II Exam Structure Continued on the Next Page...

Level II Exam Structure
General Knowledge Segment

Level II Exam General Knowledge Segment Categories	Percentage of This Segment	Number of Questions
Attributes	6%	3
File Locking	2%	1
Polyline Editing	6%	3
XYZ Filtering	2%	1
Advanced Plotting	4%	2
System Optimization	4%	2
Troubleshooting Techniques	4%	2
Advanced Utility Commands	4%	2
Loading LSP and ADS Routines	2%	1
Total	100%	50

Level II Exam Objectives

In order to successfully pass the AutoCAD Level II Certification Exam, an individual must have mastery of each of the following objectives listed below:

Objective 2.1   Productivity Techniques

Know the effects the following commands have on a drawing: Viewres, Dragmode, Regenauto, Layer-Freeze, Layer-Lock, Qtext, Vports (View Ports), Named Views, and CAL (The Geometry Calculator).

Know the purpose of Command Aliasing.

Know the following file types such as .DWG, .BAK, .LIN, .SHX, .PCP, .XLG, .DXF, .PGP, .CFG used in Intermediate AutoCAD.

Objective 2.2   User Coordinate Systems

Know the UCS command.

Know the options of the UCSICON command that how they affect the User Coordinate System.

Objective 2.3   External References

Know the advantages of using external references in a drawing.

Know all options of the Xref command.

Know the purpose and all options of the Xbind command.

Know how the Xref command handles externally referenced layers, blocks, linetypes, etc.

Objective 2.4   Model Space/Paper Space

Know the purpose of working in Model Space or Paper space.

Know the Tilemode system variable.

Know the Mview command and all of its options.

Know the Zoom-XP command for scaling to paper space units.

Know the purpose and uses of the Vplayer command on paper space viewports.

Know how to control viewport layers using the DDLMODES dialogue box.

Know the PSLTSCALE command.

Know MAXACTVP and the limitations it represents to view ports.

Level II Exam Objectives Continued on the Next Page...

Level II Exam Objectives

Objective 2.5   Advanced Grips
Know how to use grips to perform offsets and arrays using the Temporary Auxiliary Snap Grid Mode.
Know how to use the Shift key to select multiple "hot" grips.

Objective 2.6   Advanced Selection Sets
Know all entity selection modes outlined in the DDSELECT dialogue box.

Objective 2.7   Region Modeling
Know the purpose of constructing a region model.
Know the function of the Union, Subtract, and Intersection commands.
Know the Solchp command for editing a region model.
Know the purpose and uses of the DDSOLMASSP dialogue box for analyzing a region model.

Objective 2.8   System Variables
Know the following system variables:
Pickbox, Grips, Pdmode, Pdsize, Splframe, Mirrtext, Filedia, Cmddia, Savetime, and Maxsort.

Objective 2.9   Dimension Styles
Know the purpose and uses of the DDIM dialogue box.
Know how to create a dimension style.
Know how to associate dimension variables with a dimension style.
Know how to Restore a dimension style.
Know the effects Override has on a dimension style.

Objective 2.10   Object Filtering
Know the purpose of the Filter dialogue box.
Know how to create a selection set using the Filter dialogue box.

Level II Exam Objectives

Objective 2.11 Layer Filtering
Know the purpose of the Layer Filtering dialogue box which is part of the DDLMODES dialogue box.

Objective 2.12 Wildcards
Know the purpose of using various range wildcards to better query lists.

Objective 2.13 Attributes
Know the purpose of attributes associated in a drawing file.
Know the following commands: Attdef, Attdisp, Attedit, and Attext.
Know the DDATTDEF, DDATTE, and DDATTEXT dialogue boxes.

Objective 2.14 File Locking
Know the purpose of file locking.
Know the types of files produced by file locking such as .PWK, .DWK, .MNK, .SHK, etc.

Objective 2.15 Polyline Editing
Know all options of the Pedit command.
Know all Edit Vertex options of the Pedit command.
Know the effects the Ltype Gen option has on a polyline.

Objective 2.16 XYZ Filtering
Know the format for entering XYZ filters.

Objective 2.17 Advanced Plotting
Know how to create different plot parameter files (PCP).
Know the purpose and function of Freeplotting.

Level II Exam Objectives Continued on the Next Page...

Level II Exam Objectives

Objective 2.18 System Optimization
Know how to add, remove, or change a plotter in the configuration process.
Know the effects Treemax and Treedepth have on system performance.

Objective 2.19 Troubleshooting Techniques
Know how to unlock files previously locked.
Know how to recover a damaged drawing file.
Know how to redefine an existing block.
Know how to change the color of an entity by layer originally created using the Color command.

Objective 2.20 Advanced Utility Commands
Know how to create a DXF file using the DXFOUT command.
Know how to read an existing DXF file into a drawing using the DXFIN command.
Know the effects Purge has on unused items in a drawing.
Know how to use all external commands such as Catalog, Edit, Del, Dir, Sh, and Shell.

Objective 2.21 Loading LSP and ADS Routines
Know the function and uses of the Appload command.

UNIT 9

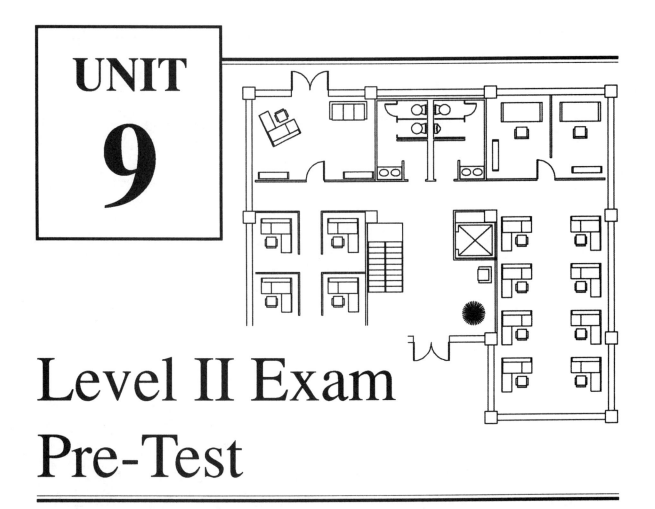

Level II Exam Pre-Test

Use this unit, Level II Exam Pre-Test, to assess your current intermediate AutoCAD skill level. The total time for this Pre-Test to be completed is 1 hour. This Level II Exam Pre-Test represents approximately 1/2 of the actual exam and is designed to highlight weak areas where more work is required to successfully pass the exam. The Level II Certification Exam Pre-Test consists of two segments:

- The first segment concentrates on drawing skills. Two problems will be drawn and the questions answered in a total of 60 minutes.

- The second segment concentrates on general AutoCAD knowledge in the form of 12 multiple choice questions. These 12 questions will be answered in 15 minutes.

Work through this Pre-Test at a good pace paying strict attention to the amount of time spent on each problem and multiple choice question. Answers for each Pre-Test question are located in Unit 13, page 226.

Notes

Level II Exam Pre-Test Section I Drawing Segment

Construct both drawing problems and answer the questions that follow each problem. The problems may be completed in any order. You should allow yourself a total of 60 minutes to complete both problems.

When both problems have been completed and time still remains, use the extra time to carefully check your answers.

Problem 1

Gasket5.Dwg

Directions for Gasket5.Dwg

Start a new drawing called Gasket5. Even though this is a metric drawing, no special limits need to be set. Keep the default setting of decimal units but change the number of decimal places past the zero from 4 to 0, (Zero). Begin constructing Gasket5 with the center of the 38 diameter hole at absolute coordinate (114,116). Dimensions do not have to be added to this drawing. Answer the questions on the next page regarding this drawing.

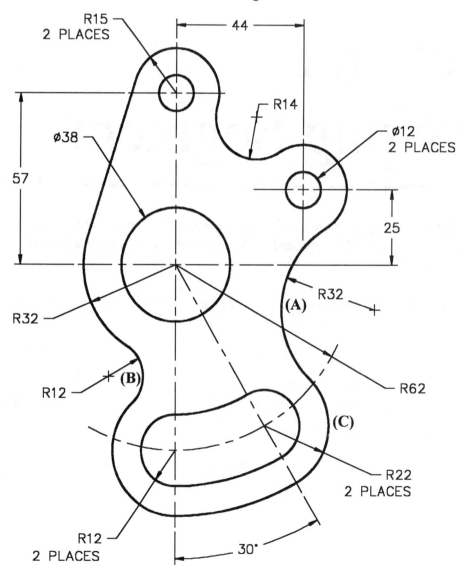

Questions for Gasket5.Dwg

Refer to the drawing of Gasket5 on the previous page to answer questions #1 through #5:

1. The total surface area of Gasket5 with all holes and slots removed is closest to
   - (A) 6683 units.
   - (B) 6685 units.
   - (C) 6687 units.
   - (D) 6689 units.
   - (E) 6691 units.

2. The total distance from the center of the 32 radius arc "A" to the center of the 12 radius arc "B" is closest to
   - (A) 93
   - (B) 95
   - (C) 97
   - (D) 99
   - (E) 101

3. The total length of the 22 radius arc "C" is
   - (A) 33
   - (B) 35
   - (C) 37
   - (D) 39
   - (E) 41

4. The absolute coordinate value of the centroid of Gasket5 is located closest to
   - (A) 123,104
   - (B) 125,106
   - (C) 127,108
   - (D) 125,110
   - (E) 125,112

5. Change both 12 diameter holes to a new diameter of 20 units. Also change the 38 diameter hole to a new diameter of 45 units. The new absolute coordinate value of the centroid of Gasket5 is closest to
   - (A) 125,107
   - (B) 125,109
   - (C) 127,111
   - (D) 125,113
   - (E) 127,115

Provide the answers in the spaces below:

1. _____

2. _____

3. _____

4. _____

5. _____

CONTINUE ON TO THE NEXT PAGE...

Problem 2

Fdn.Dwg

Directions for Fdn.Dwg

Start a new drawing called Fdn. Change the default units from decimal to architectural. Do not add dimensions to this drawing. Unless otherwise noted, the thicknesses of all concrete block walls is 8" and the thicknesses of all footings is 16" as in the illustration below. Answer the questions on the next page regarding this drawing.

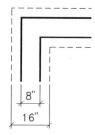

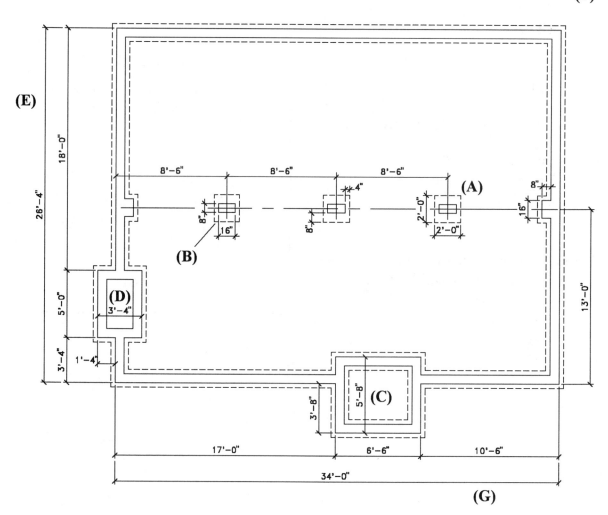

Questions for Fdn.Dwg

Refer to the drawing of Fdn on the previous page to answer questions #6 through #10:

6. The total area of all concrete block regions including the three piers located in the center of the foundation plan is closest to
   - (A) 91 sq. ft.
   - (B) 93 sq. ft.
   - (C) 95 sq. ft.
   - (D) 97 sq. ft.
   - (E) 99 sq. ft.

7. The total distance from the intersection of the footing at "A" to the intersection of the footing at "B" is closest to
   - (A) 13 ft.
   - (B) 15 ft.
   - (C) 17 ft.
   - (D) 19 ft.
   - (E) 21 ft.

8. The total area of all footing regions including the three footings that support the piers located in the center of the foundation plan is closest to
   - (A) 206 sq. ft.
   - (B) 208 sq. ft.
   - (C) 210 sq. ft.
   - (D) 212 sq. ft.
   - (E) 214 sq. ft.

9. The total distance from the middle of the concrete slab area at "C" to the middle of the fireplace area at "D" is closest to
   - (A) 15 ft.
   - (B) 7 ft.
   - (C) 19 ft.
   - (D) 21 ft.
   - (E) 23 ft.

10. Increase the size of the foundation plan using the Stretch command according to the following specifications:
    - (a) Use a crossing box from "E" to "F" to stretch the foundation plan straight up at a distance of 2'-6".
    - (b) Use a crossing box from "G" to "F" to stretch the foundation plan directly to the left at a distance of 18".

    The new area of the concrete block region **not including the three piers** located at the middle of the foundation plan is closest to
    - (A) 94 sq. ft.
    - (B) 96 sq. ft.
    - (C) 98 sq. ft.
    - (D) 100 sq. ft.
    - (E) 102 sq. ft.

Provide the answers in the spaces below:

6._____

7._____

8._____

9._____

10._____

END OF SECTION I - DO NOT PROCEED FURTHER UNTIL TOLD TO DO SO

Notes

Level II Exam Pre-Test Section II General Knowledge Segment

Answer each of the 12 multiple-choice questions. The questions may be answered in any order. It is considered good practice to answer the easier questions first. If a question seems difficult, do not waste time trying to answer it. Go on to the next question and come back to the difficult question or questions later. Be sure to provide the best possible answer for each question.

You should allow yourself a total of 15 minutes to answer all 12 multiple-choice questions.

Place the best answer in the appropriate box for each of the following questions. Unless otherwise specified, all questions are "Single Answer Multiple Choice".

Question 11 is considered a "Multiple-Answer Multiple-Choice question. Supply all possible answers that satisfy the particular question.

11. Of the following operations, those that will speed up the regeneration of a drawing include
    (A) setting VIEWRES from 100 to 500.
    (B) turning QTEXT On.
    (C) freezing unused layers.
    (D) turning FILL Off.
    (E) turning DRAGMODE On.

12. If an externally referenced drawing called "House.Dwg" contains a layer called "FND", the name of the layer in the current drawing will be
    (A) FND.
    (B) HOUSE-FND.
    (C) HOUSE|FND.
    (D) HOUSE/FND.
    (E) Either A or B.

Question 13 is considered a "Multiple-Answer Multiple-Choice question. Supply all possible answers that satisfy the particular question.

13. Valid dimension subcommands used to manage dimension styles include
    (A) SAVE.
    (B) END.
    (C) RESTORE.
    (D) VARIABLES.
    (E) HORIZONTAL.

14. To scale the entities contained in Paper Space viewports, use the ZOOM command along with the
    (A) XP option.
    (B) Extents option.
    (C) Window option.
    (D) Dynamic option.
    (E) Vmax option.

15. Grips will be used to move a vertical dimension closer to the drawing. As the dimension is selected, grips appear and the dimension entity highlights. Making the grip at the text location "hot" and stretching the dimension to a new location
    (A) stretches the entire dimension closer to the object.
    (B) stretches only the dimension line and arrows closer to the object.
    (C) stretches only the dimension text closer to the object.
    (D) shortens the length of the extension lines.
    (E) None of the above.

16. The command used to edit primitives belonging to a region model is
    (A) EDIT.
    (B) REDIT.
    (C) MODIFY
    (D) CHANGE
    (E) SOLCHP

Question 17 is considered a "Short Answer or Fill In the Blank" question. Provide the correct answer in the space provided that satisfies the particular question.

17. The command used to create an attribute is

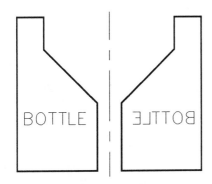

18. In the illustration above, the system variable that controls whether text is reflected backwards or is reflected but still readable is

(A) MIRRTEXT
(B) REFLTEXT
(C) MIRRORTX
(D) REFLCTX
(E) FLIPTEXT

19. The command used to place blocks along an irregular shaped entity at a user specified distance is

(A) ARRAY
(B) MEASURE
(C) DIVIDE
(D) Only B and C
(E) None of the above

Given the following block names, answer the next question by supplying the appropriate range wildcard:

CHAIR-1
CHAIR-2
CHAIR-3
CHAIR-10
CHAIR-20
CHAIR-30
COUCH-A
COUCH-B

20. To produce a list of all blocks that begin with "CHAIR-" and are followed by a double digit, use the following range wildcard:

(A) CHAIR##
(B) CHAIR-[##]
(C) CHAIR-##
(D) ##CHAIR
(E) [##]-CHAIR

Question 21 is considered a "Multiple-Answer Multiple-Choice question. Supply all possible answers that satisfy the particular question.

21. Valid options of the Edit Vertex subcommand of PEDIT include

(A) Trim.
(B) Move.
(C) Scale.
(D) Copy.
(E) Break.

```
┌─────────────────────────────────────────┐
│      Entity Selection Settings           │
│  Selection Modes                         │
│  ┌─────────────────────────────────────┐ │
│  │ ⊠ Noun/Verb Selection               │ │
│  │ ☐ Use Shift to Add                  │ │
│  │ ☐ Press and Drag                    │ │
│  │ ⊠ Implied Windowing                 │ │
│  │ ┌─────────────────────────────────┐ │ │
│  │ │  Default Selection Mode         │ │ │
│  │ └─────────────────────────────────┘ │ │
│  └─────────────────────────────────────┘ │
│  Pickbox Size                            │
│  ┌─────────────────────────────────────┐ │
│  │   Min        Max                    │ │
│  │                              ☐      │ │
│  │  ◄├──┤─────────────────►            │ │
│  └─────────────────────────────────────┘ │
│      ┌─────────────────────────────┐     │
│      │  Entity Sort Method...      │     │
│      └─────────────────────────────┘     │
│  ┌────────┐  ┌────────┐  ┌─────────┐     │
│  │   OK   │  │ Cancel │  │ Help... │     │
│  └────────┘  └────────┘  └─────────┘     │
└─────────────────────────────────────────┘
```

22. In the illustration at the right of the Entity Selection Settings dialogue box and given the default settings, which of the following is true:

   (A) The individual has to first enter an edit command before selecting the entities to edit.

   (B) The individual has to hold down the shift key when picking entities to add to a selection set.

   (C) The individual has to hold down the pick button when selecting a window or crossing box since selection drag mode is activated.

   (D) The individual has the option of picking an entity which will automatically activate window or crossing mode.

   (E) The individual has the option of first selecting entities at the "Command:" prompt and then enter the desired edit command.

END OF SECTION II

UNIT
10

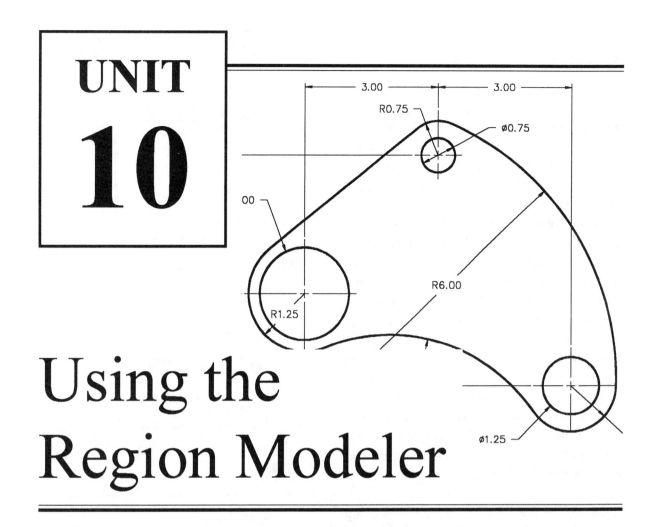

Using the
Region Modeler

Use this unit to learn the mechanics of constructing a region model of an object and for what purpose. This unit begins with a basic introduction of the region modeler and how to load the modeler into a drawing file followed by the methods of setting up a region model. Basic region modeling construction commands are explained on how they differ from conventional AutoCAD drawing and editing commands. The strength of the region modeler is explained in its ability to provide an analysis of the region model consisting of area, perimeter, bounding box, centroid, moments of intertia, and other analysis functions. This unit concludes with a tutorial exercise taking the individual on a step by step method of first constructing a region model, performing an analysis of the region, editing certain primitives that belong to the region, and performing another analysis of the edited region.

Introducing the Region Modeler

As an alternate method of performing geometric constructions, a region modeler has been developed. A region model represents a closed two-dimensional shape that is treated as a single entity similar to a block. As a result, such properties as Area and Perimeter are associated with regions. In some cases, the use of a region modeler may make geometric constructions easier and less time consuming. The method of constructing regions that makes it completely different than conventional construction methods is the use of "boolean operations". These operations of Union, Subtraction, and Intersection allow entities to be joined together into a single entity or subtracted from each other leaving a difference in both entities. Once a region is constructed, a separate mass property utility is available to perform calculations such as area, perimeter, centroid, and even moments of inertia.

Illustrated at the right is a method of loading the region modeler using the Appload command.

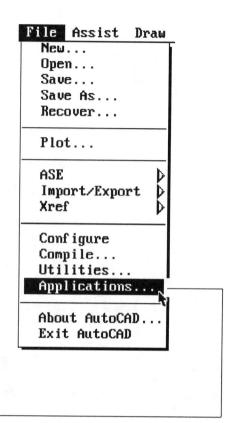

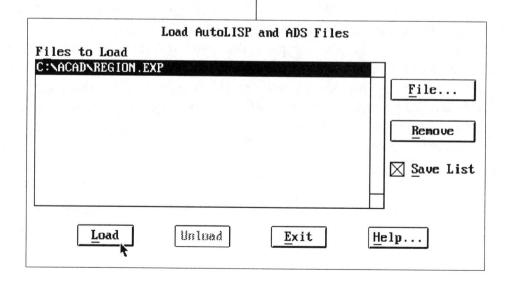

Introducing the Region Modeler

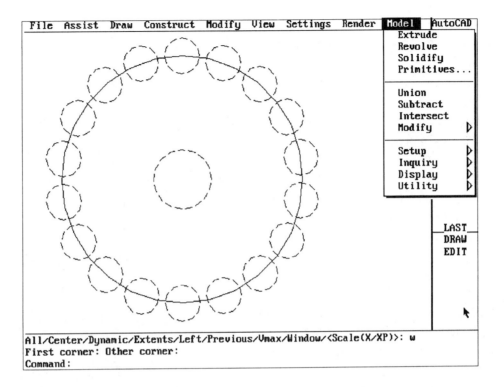

Illustrated above and to the right is one area used to obtain region modeling commands, namely the pulldown menu area. To display the dialogue box illustrated at the right, pick "Model" from the pulldown menu. As this menu displays, notice some commands appear more bolder than others. The commands identified in bold print are valid commands that can be used to create, edit, or analyze the region model. Commands that are greyed out are supported only the the Advanced Modeling Extension used to create 3-dimensional solid models.

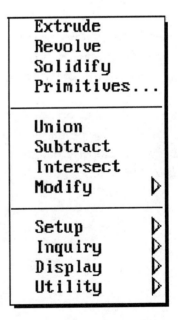

Setting Up a Region Model

From the "Setup" area of the "Model" dialogue box activates the "Variables" submenu. Selecting "Variables" in turn displays the Region Modeler System Variables dialogue box illustrated below. Use this dialogue box to set values affecting the region modeler. This dialogue box may also be displayed by entering DDSOLVAR at the command prompt from the keyboard. Areas of the dialogue box in bold print are active areas supported by the region modeler. Areas that are greyed out support only the 3-dimensional Advanced Modeling Extension.

For the purposes of this section, the following areas of this dialogue box will be explained; Selecting "Units . . ." activates another dialogue box allowing the individual to make changes to the currents units used to create the region model; Selecting "Hatch Patterns . . ." activates a dialogue box allowing the individual to use different hatch patterns to be used when the region is automatically crosshatched. Both of these dialogue boxes will be discussed in greater detail on the next page. The "Other Parameters . . ." dialogue box pertains mainly to system variables that primarily affect the 3-dimensional solid model created in the Advanced Modeling Extension.

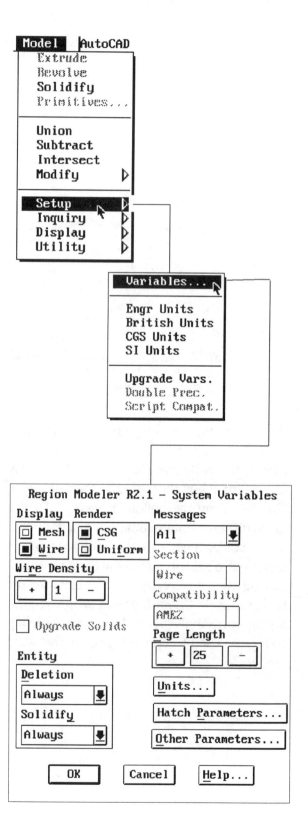

Setting Up a Region Model

Selecting the "Units . . ." button from the Region Modeler System Variables dialogue box displays the current units used when calculations are performed. Two measurements may be changed to reflect the desired units of the region model, namely length and Area. The remaining two measurements of volume and mass are reserved for a 3-dimensional solid model constructed using the Advanced Modeling Extension. The default units for all region models are centimeters for length and square centimeters for area.

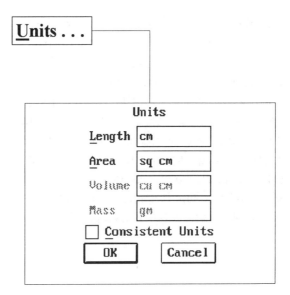

When changing from centimeters to other forms of units, enter "in" as the length unit value in place of "cm" as in the example at the right. To have the area reflect these new units, pick the check box labeled "Consistent Units" at the right. This will automatically update all units to the current value placed in the "Length" edit box.

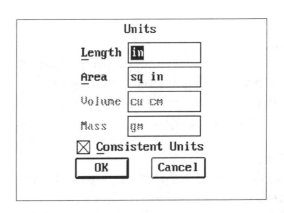

Selecting "Hatch Parameters . . ." from the "Region Modeler System Variables" dialogue box displays the current hatch parameters. These parameters will be used to automatically hatch a region as it is being created. All supplied hatch patterns are supported in the region modeler. If the pattern "None" is substituted for the current pattern "U", no hatch pattern is applied to the region during the creation, editing or analysis processes.

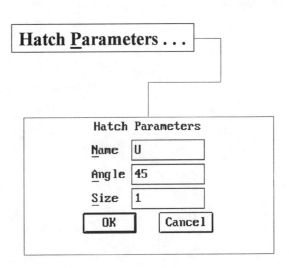

Region Modeling Construction Techniques

One of the most fundamental methods of creating regions is through the use of Constructive Solids Geometry. This is accomplished through the boolean operations of union, subtraction, and intersection. All three operations are explained below and utilize the circle and rectangle illustrated at the right. To aid in the construction process, the four lines representing the rectangle have been converted into a single polyline. The Rectang command automatically draws a rectangle as a polyline.

Individual Entities

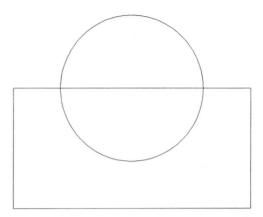

Union
Use this command to join a group of entities into one complete region. Selecting the circle and rectangular polyline creates the region illustrated at the right.

Command: **Union**
Select objects: *(Select the circle and rectangle)*
Select objects: *(Strike Enter to create a union of both entities)*

Creating a Region by Union

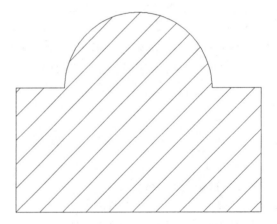

Region Modeling Construction Techniques

Subtraction

Use this command to subtract an entity or series of entities from a source object. To create the cove illustrated at the right, select the rectangle as the source object; the circle is subtracted from the rectangle to form the cove shape.

Command: **Subtract**
Source objects...
Select objects: *(Select the rectangle as the source object)*

Creating a Region by Subtraction

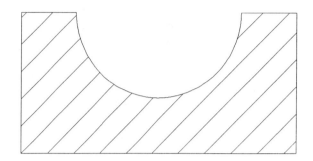

Intersection

Use this command to produce the region from the intersection of two or more entities. The region shape illustrated at the right is common to both the circle and rectangle.

Command: **Solint**
Select objects: *(Select the circle and rectangle)*
Select objects: *(Strike Enter to create the intersection)*

Creating a Region by Intersection

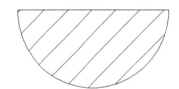

Analyzing a Region Model

Selecting "Mass Property . . ." from the "Inquiry" submenu of the "Model" pulldown menu area activates the "Mass Properties" dialogue box illustrated below. This dialogue displays various calculations based on a selected region. The following properties of the region selected are calculated and listed below:

Area

Perimeter

Bounding Box

Centroid

Moments of Inertia

Product of Inertia

Radii of Gyration

Principal Moments and X-Y directions about centroid

The "Mass Properties" dialogue box may also be displayed by entering the command DDSOLMASSP from the keyboard. If Solmassp is entered from the keyboard and a region is selected, the mass properties are displayed in a text screen format.

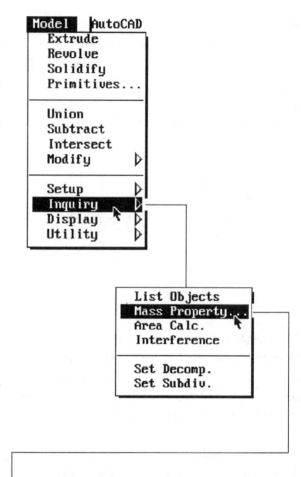

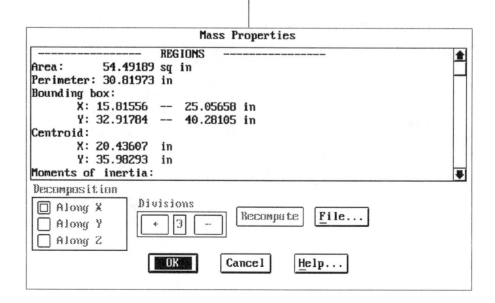

Analyzing a Region Model

The following mass property calculations are explained below:

The Area option displays the enclosed area of the region.

The Perimeter represents the total length of the inside or outside of the region.

The Bounding Box is represented by a rectangular box that totally encloses the region. The box is identified by two diagonal points as in "A" and "B" in the illustration at the right.

The Centroid represents the center of the regions area. The location of the centroid is identified by a point entity placed on the current layer. The System Variable PDMODE controls the style of point displayed.

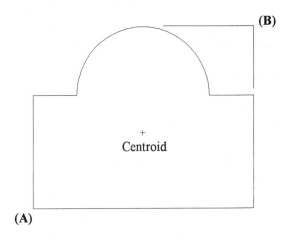

If the "File . . ." button is selected from the "Mass Properties" dialogue box, the following prompt appears:
   Write to a file? <N>:

Answering "Yes" displays the next prompt:
   File name <Default drawing name>:

After entering a file name, all mass property calculations are written out to a DOS text file similar to the illustration at the right. The text file that is created has the extension .Mpr.

Area:	54.49189 sq cm
Perimeter:	30.81973 cm
Bounding box:	X: 46.34819 — 55.58921 cm
	Y: 0.8098652 — 8.173072 cm
Centroid:	X: 50.9687 cm
	Y: 3.874955 cm
Moments of inertia:	X: 1012.65 sq cm sq cm
	Y: 141884.2 sq cm sq cm
Product of inertia:	XY: 10762.22 sq cm sq cm
Radii of gyration:	X: 4.310859 cm
	Y: 51.02713 cm

Principal moments(sq cm sq cm) and X-Y directions about centroid:
   I: 194.4393 about [1 0]
   J: 324.771 about [0 1]

Editing a Region Model

The spoke assembly illustrated at the right was constructed as a region; it is now required to go through a design change; the 8.000 unit diameter hole in the center of the spoke needs to be changed to a new diameter of 4.000 units. Under normal circumstances, the Change command could have been used to change a circle to a new radius or diameter. However since the 8.000 diameter circle is considered a primitive component of the total region, it is difficult to select just the circle using convential editing commands. As a result, the Solchp command (Solid Change Property) is reserved for editing primitive shapes that belong to a region. Use this command to change the size or color of a primitive; you can also delect, copy, move, or replace a primitive.

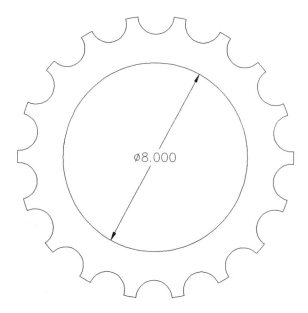

Editing a Region Model

In the illustration at the right, the Solchp command is used to change the size of the large hole. After selecting the entire region followed by the desired primitive to change (the large circle at "A"), use the Size option of the command. Notice the Motion Control System icon (MCS) appear giving you directions for such options as Move. After changing the size of the large hole, the icon disappears and the change is made.

Command: **Solchp**
Select a region: *(Select any part if the spoke illustrated at the right)*
Select primitive: *(Select the large circle at "A")*
Color/Delete/Evaluate/Instance/Move/Next/Pick/Replace/Size/eXit <N>: **Size** *(The Motion Control System icon appears illustrated at the right)*
Radius of circle <4>: **2.000**
Color/Delete/Evaluate/Instance/Move/Next/Pick/Replace/Size/eXit <N>: **X** *(To exit this command)*

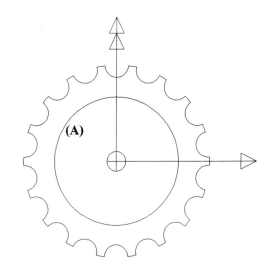

When the Solchp command is properly exited, the region updates to the latest changes such as the size of the large center circle changing from 8.000 to 4.000 units in diameter.

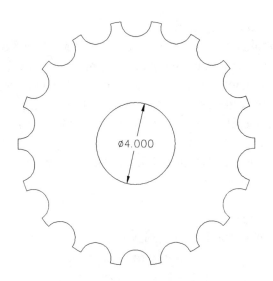

Tutorial Exercise #4
Gasket.Dwg

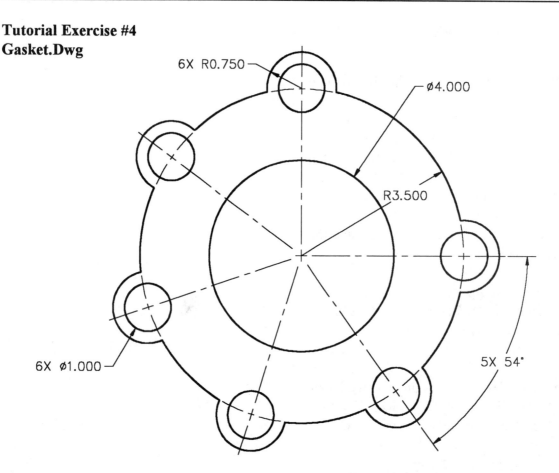

6X R0.750

ø4.000

R3.500

6X ø1.000

5X 54°

Purpose:
This tutorial is designed to construct the Gasket above using Region Modeling techniques. Once the object is constructed, it will be analyzed for accuracy.

System Settings:
Begin a new drawing called "Gasket". Use the Units command to change the number of decimal places past the zero from 4 to 2. Keep the remaining default unit values. Keep the current limits set to 0,0 for the lower left corner and 12,9 for the upper right corner. The Grid or Snap commands fo not need to be set to any certain values.

Layers:
Create the following layers with the format:
Name-Color-Linetype
Object - Green - Continuous

Suggested Commands:
Begin by locating the center of the large 4.000 diameter hole at absolute coordinate 6.000,4.500. Construct all necessary circles; use the Array command to crate the circular pattern of 6 holes spaced 54 degrees away from each other. Begin joining the outer perimeter of the Gasket using the Union command. To create the holes as cut-outs, use the Subtract command. Use the Solmassp or DDSOLMASSP commands to analyze the region. Edit the region by changing the size of the large inside hole and the six holes arranged in the circular pattern. Perform another analysis of the region.

Step #1

Construct two circles, one of radius 3.500 and the other of diameter 4.000. Use the center point of 6.000,4.500 for both circles. The "@" symbol is used for the center of the second circle to use the last known coordinate as the center.

Command: **Circle**
3P/2P/TTR/<Center point>: **6.000,4.500**
Diameter/<Radius>: **3.500**

Command: **Circle**
3P/2P/TTR/<Center point>: **@**
Diameter/<Radius>: **Diameter**
Diameter: **4.000**

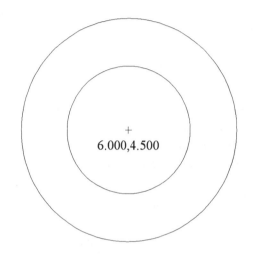

Step #2

Begin forming the lobes and holes of the Gasket by constructing a circle of radius 0.750 and circle of diameter 1.00 from the upper quadrant of the large circle at "A" illustrated at the right.

Command: **Circle**
3P/2P/TTR/<Center point>: **Qua**
of *(Select the quadrant of the large circle at "A")*
Diameter/<Radius>: **0.750**

Command: **Circle**
3P/2P/TTR/<Center point>: **@**
Diameter/<Radius>: **Diameter**
Diameter: **1.000**

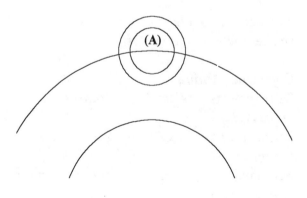

Step #3

Use the Array command to duplicate the last two circles six times in an included angle of 270 degrees. Since this is a positive angle, the direction of the array will be in the counter-clockwise direction.

Command: **Array**
Select objects: *(Select the last two circles)*
Select objects: *(Strike Enter to continue)*
Rectangular or Polar array (R/P) <R>: **Polar**
Center point of array: **Cen**
of *(Select the center of the circle at "A")*
Number of items: **6**
Angle between items (+=CCW, -=CW): **270**
Rotate object as they are copied? <Y>: *(Strike Enter to accept this default value)*

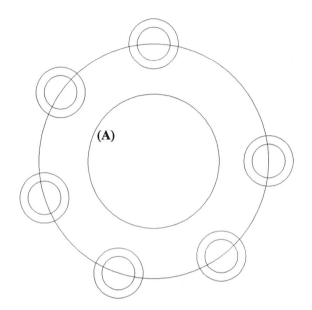

Step #4

Begin creating the region by using the Union command to join the large circle with the six smaller 0.750 radius circles.

Command: **Union**
Select objects: *(Select the large dashed circle at the right)*
Select objects: *(Select the six dashed circles at the right)*
Select objects: *(Strike Enter to create a union of both entities)*

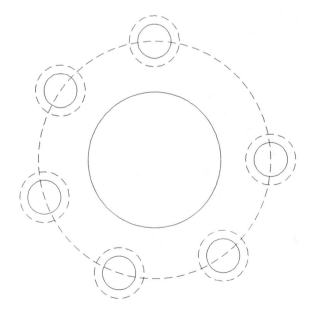

Step #5

The result of the Union command is illustrated at the right. The smaller circles blend in with the large circle to form a region identified by the crosshatching pattern. This pattern is controlled by the Solhpat variable. The spacing of the hatch pattern is controlled by the Solhsize variable. Finally the angle of the hatch pattern is controlled by the Solhangle variable. If no hatch pattern is desireable, set the Solhpat variable to none. Notice that the hatch pattern goes through all holes. Follow the next step to subtract these holes from the original region to form a new region.

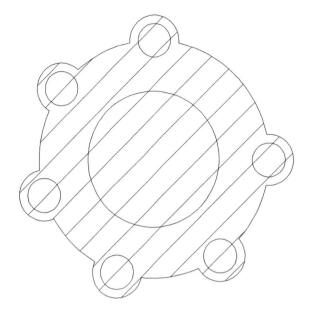

Step #6

Use the Subtract command to subtract the seven dashed holes at the right from the region.

Command: **Subtract**
Source objects...
Select objects: *(Select the region at "A" as the source object)*
Select objects: *(Strike Enter to continue)*
Objects to subtract from them...
Select objects: *(Select all seven dashed circles illustrated at the right. This will subtract them from the original region)*

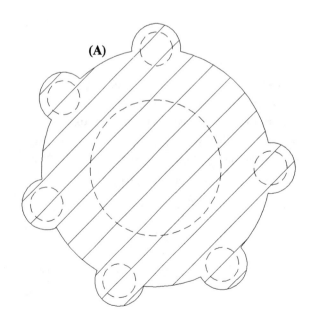

(A)

Step #7

As the holes subtract from the first region, a new region illustrated at the right is formed. The same hatch pattern, hatch size, and hatch angle are used on the new region.

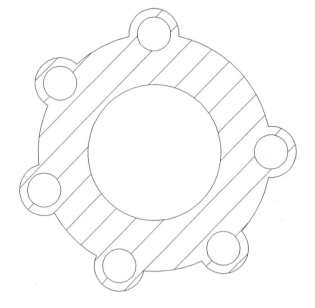

Step #8

Setting the hatch pattern to none displays the region similar to the illustration at the right. Once the hatch pattern is displayed and the Solhpat variable is set to none, boolean operations or solid editing operations update the region to the new hatch pattern. The use of the Regen command is uneffective.

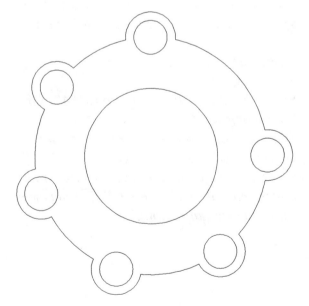

Step #9

To perform such calculations on a region, the DDSOLMASSP or Solmassp commands are used. These commands are very useful to extract the area, perimeter, bounding box, and centroid of a region. Other information such as Moments of inertia, Product of inertia, Radii of gyration, and Principal moments are also calculated. The differences in the two commands lie in the methods they display this information. The DDSOLMASSP command displays all calculations in a dialogue box illustrated just below. The Solmassp command provides the same information about the region; it displays its information in text form. Once the mass properties of a region are found, the centroid is automatically marked by the current point mode. This is displayed as the "+" sign at the right signifying a Pdmode of 2.

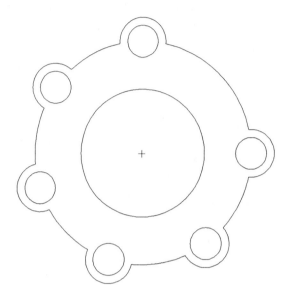

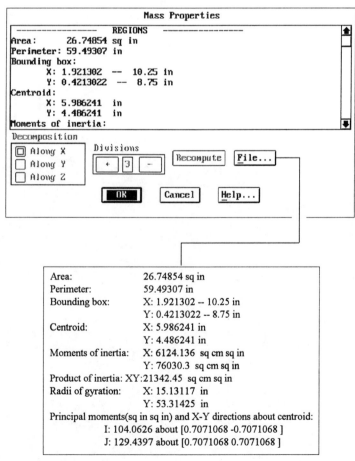

Step #10

A change has occurred in the original design of the Gasket. The large 4.000 diameter hole needs to be changed to a new diameter of 5.000 units. The Solchp command along with the Size option will be used to perform this task.

Command: **Solchp**
Select a region: *(Select the region at "A")*
Select primitive: *(Select the 4.000 unit diameter hole at "B")*
Color/Delete/Evaluate/Instance/Move/Next/ Pick/Replace/Size/eXit <N>: **Size** *(The Motion Control System icon will appear)*
Radius of circle <2>: **2.500**
Color/Delete/Evaluate/Instance/Move/Next/ Pick/Replace/Size/eXit <N>: **X** *(To exit this command)*

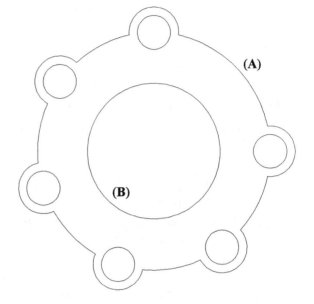

Step #11

Notice at the right that the 4.000 diameter hole is changed to the new diameter of 5.000 when the Solchp command is properly exited. Another change must be made to the Gasket; all six 1.000 diameter holes must be changed to a new diameter of 0.750. The Solchp command is first made to change one hole. Then a copy or instance of the hole is made before it is arrayed and used to replace all existing holes. Follow the prompts below to perform this step:

Command: **Solchp**
Select a region: *(Select the region at "A")*
Select primitive: *(Select the 1.000 unit diameter hole at "B")*
Color/Delete/Evaluate/Instance/Move/Next/Pick/Replace/Size/eXit <N>: **Size** *(The Motion Control System icon will appear)*
Radius of circle <0.5>: **0.375**
Color/Delete/Evaluate/Instance/Move/Next/Pick/Replace/Size/eXit <N>: **Instance** *(To create a copy of the new primitive)*
Color/Delete/Evaluate/Instance/Move/Next/Pick/Replace/Size/eXit <N>: **X** *(To exit this command)*

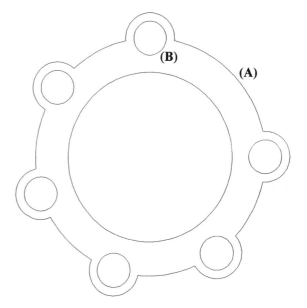

After performing the Instance option of the Solchp command, two holes now exist at location "B".

Step #12

Use the Array command to copy the last entity six times in an included angle of 270 degrees. The Last option is used to select the hole; otherwise use the Window option to select only the copy of the hole at "A". After performing the array, each circle represents six individual regions which will replace the existing 1.000 diameter holes. Follow the prompts below to perform this operation:

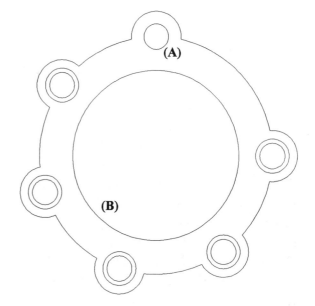

Command: **Array**
Select objects: **L** *(To select the last circle at "A" which was copied using the Instance option of Solchp)*
Select objects: *(Strike Enter to continue)*
Rectangular or Polar array (R/P) <R>: **Polar**
Center point of array: **Cen**
of *(Select the center of the circle at "B")*
Number of items: **6**
Angle between items (+=CCW, -=CW): **270**
Rotate object as they are copied? <Y>: *(Strike Enter to accept this default value)*

Step #13

Rather than change the individual size of each
1.000 diameter hole, the Replace option of the
Solchp command will be used.

Command: **Solchp**
Select a region: *(Select the region at "A")*
Select primitive: *(Select the 1.000 unit diam-
eter hole at "B")*
Color/Delete/Evaluate/Instance/Move/Next/
Pick/Replace/Size/eXit <N>: **Replace**
Select region to replace primitive: *(Select the
inside region at "C")*
Retain detached primitive? <N>: *(Strike Enter
to accept this default value)*
Color/Delete/Evaluate/Instance/Move/Next/
Pick/Replace/Size/eXit <N>: **Pick**
Select primitive: *(Select the next 1.000 diam-
eter hole)*

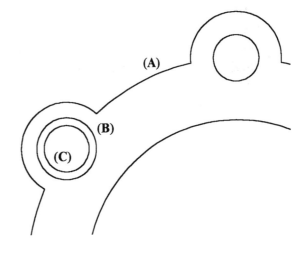

Continue using the Replace option of Solchp
until all 1.000 diameter holes at "A", "B", "C",
and "D" are changed to the new diameter of
0.750. Exit the Solchp command to update all
hole changes.

Color/Delete/Evaluate/Instance/Move/Next/
Pick/Replace/Size/eXit <N>: **X** *(To exit this
command)*

Two holes still exist at "E". Use the Erase
command and the Window option to window
in the hole to erase it. Selecting the hole may
select the entire region.

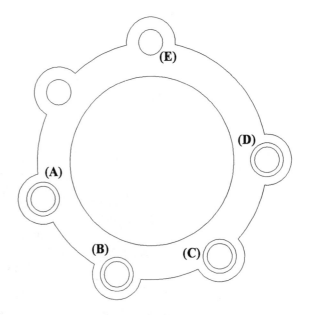

Step #14

Use the DDSOLMASSP command to perform a mass property calculation on the new region. Compare the differences in the Area, Perimeter, and Centroid in the dialogue box below with the results located in the dialogue box on page 179 on the region before the changes were made. Create a text file of the mass property calculations similar to the illustration at the bottom of this page.

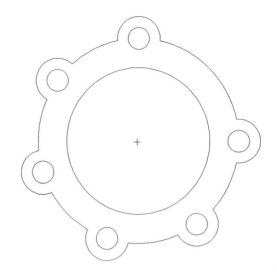

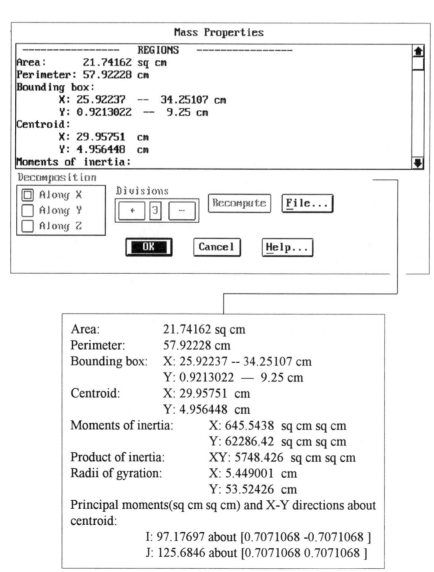

Area: 21.74162 sq cm
Perimeter: 57.92228 cm
Bounding box: X: 25.92237 -- 34.25107 cm
                 Y: 0.9213022 — 9.25 cm
Centroid: X: 29.95751 cm
                 Y: 4.956448 cm
Moments of inertia: X: 645.5438 sq cm sq cm
                         Y: 62286.42 sq cm sq cm
Product of inertia: XY: 5748.426 sq cm sq cm
Radii of gyration: X: 5.449001 cm
                         Y: 53.52426 cm
Principal moments(sq cm sq cm) and X-Y directions about centroid:
                 I: 97.17697 about [0.7071068 -0.7071068]
                 J: 125.6846 about [0.7071068 0.7071068]

Step #15

The complete region after editing is illustrated
below. Dimensions may be added to the model
at a later date.

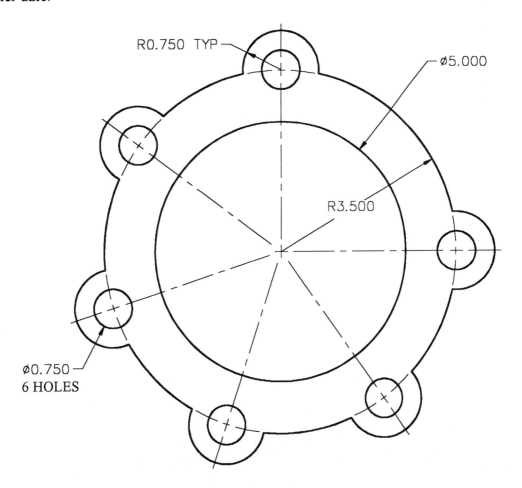

Notes

UNIT 11

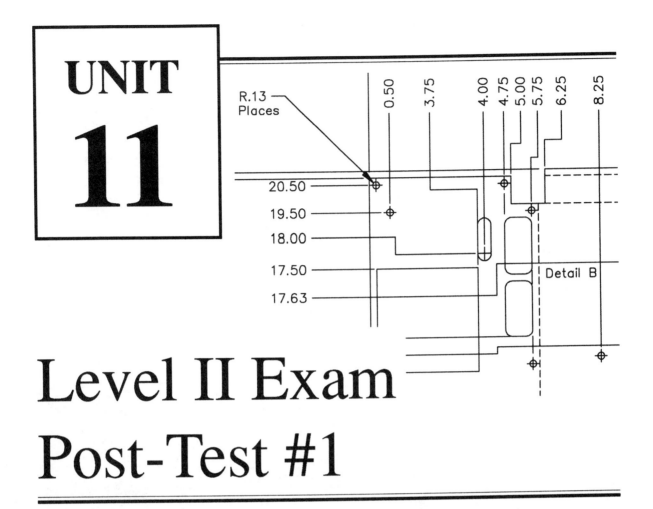

Level II Exam Post-Test #1

Use the Level II Certification Exam Post-Test #1 for more practice in preparing for the full 3-hour AutoCAD Level II Certification Exam. This test is designed to be completed in 2 hours, which represents 2/3 of the actual exam. The AutoCAD Level II Certification Exam Post-Test #1 consists of the following two segments:

- The first segment concentrates on drawing skills. Three problems will be drawn and the questions answered in a total of 90 minutes.

- The second segment concentrates on general AutoCAD knowledge in the form of 25 general knowledge questions. These 25 questions are to be answered in 30 minutes.

Work through the Level II Post-Test #1 at a good pace paying strict attention to the amount of time spent on each drawing problem and general knowledge question. Answers for all Level II Post-Test #1 questions are located in Unit 13, page 227.

Notes

Level II Exam Post-Test #1 Section I Drawing Segment

Construct the three drawing problems assigned to this segment and answer the questions that follow each problem. The problems may be completed in any order. You should allow yourself a total of 90 minutes to complete all three problems.

When all problems have been completed and time still remains, use the extra time to carefully check your answers.

Problem 1

Bldg1.Dwg

Directions for Bldg1.Dwg

Start a new drawing called Bldg1. Change from decimal units to architectural units. Keep all remaining default values. **All block wall thicknesses identified by crosshatching measure 12". All other interior wall thicknesses measure 6".** Do not add any dimensions to this drawing. Answer the questions on the next page regarding this drawing.

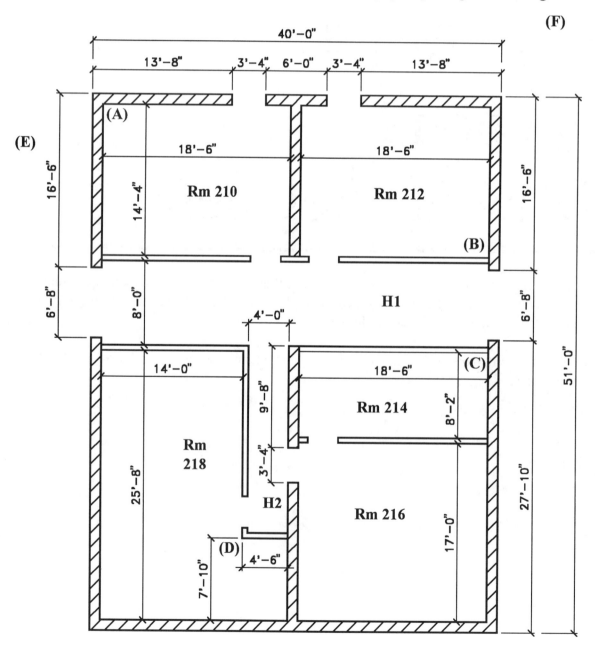

Questions for Bldg1.Dwg

Refer to the drawing of Bldg1 on the previous page to answer questions #1 through #5 below:

1. The total area of all concrete block walls identified by the crosshatching pattern is closest to
    - (A) 183 sq. ft.
    - (B) 186 sq. ft.
    - (C) 189 sq. ft.
    - (D) 192 sq. ft.
    - (E) 195 sq. ft.

2. The total area of rooms 210, 216, and 218 is closest to
    - (A) 971 sq. ft.
    - (B) 974 sq. ft.
    - (C) 977 sq. ft.
    - (D) 980 sq. ft.
    - (E) 983 sq. ft.

3. The angle formed in the XY plane from the inside corner intersection at "A" to the inside corner intersection at "B" is closest to
    - (A) 333 degrees.
    - (B) 335 degrees.
    - (C) 337 degrees.
    - (D) 339 degrees.
    - (E) 341 degrees.

4. The delta X,Y distance from the inside corner intersection at "C" to the outside intersection of the hallway corner at "D" is closest to
    - (A) -24'-0",-17'-10"
    - (B) -24'-0",-18'-1"
    - (C) -24'-0",-18'-4"
    - (D) -24'-0",-18'-7"
    - (E) -24'-0",-18'-10"

5. Stretch the building straight up using a crossing box from "E" to "F" and at a distance of 5'-4". The total area of rooms 212, 214, and 218 is closest to
    - (A) 900 sq. ft.
    - (B) 903 sq. ft.
    - (C) 906 sq. ft.
    - (D) 909 sq. ft.
    - (E) 912 sq. ft.

Provide the answers in the spaces below:

1._____

2._____

3._____

4._____

5._____

CONTINUE ON TO THE NEXT PAGE...

Problem 2

Gusset.Dwg

Directions for Gusset.Dwg

Start a new drawing called Gusset. Begin the construction of the Gusset illustrated below by keeping the default units set to decimal but change the number of decimal places past the zero from 4 to 2. No special limits need be set for this drawing. Do not add any dimensions to this drawing. Answer the questions on the next page regarding this drawing.

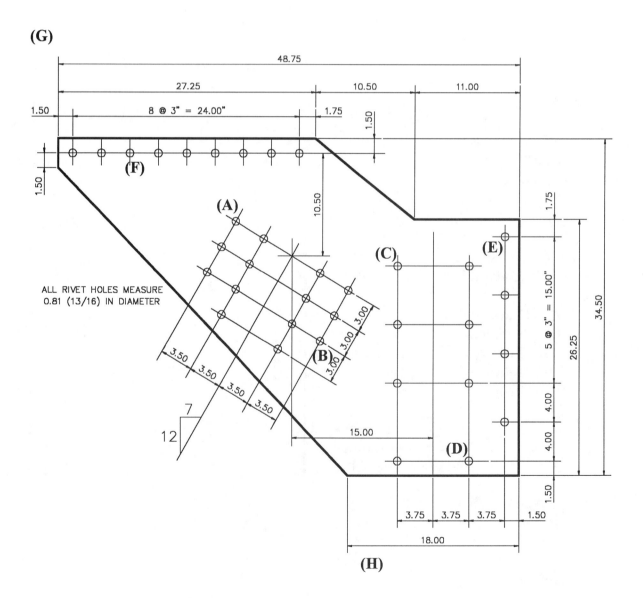

Questions for Gusset.Dwg

Refer to the drawing of Gusset on the previous page to answer questions #6 through #10 below:

6. The total area of the Gusset plate with all 35 rivet holes removed is closest to
    - (A) 1045.46
    - (B) 1050.67
    - (C) 1055.21
    - (D) 1060.40
    - (E) 1065.88

7. The delta X,Y distance from the center of hole "A" to the center of hole "B" is
    - (A) 9.07,-12.19
    - (B) 9.07,-12.24
    - (C) 9.07,-12.29
    - (D) 9.07,-12.34
    - (E) 9.07,-12.39

8. The angle formed in the XY plane from the center of hole "C" to the center of hole "D" is
    - (A) 285 degrees.
    - (B) 288 degrees.
    - (C) 291 degrees.
    - (D) 294 degrees
    - (E) 297 degrees.

9. The angle formed in the X-Y plane from the center of hole "E" to the center of hole "F" is
    - (A) 165 degrees.
    - (B) 168 degrees.
    - (C) 171 degrees.
    - (D) 174 degrees.
    - (E) 177 degrees.

10. Stretch the Gusset plate directly to the left using a crossing box from "G" to "H" and at a distance of 3.00 units. The new area of the Gusset plate with all 35 rivet holes removed is
    - (A) 1121.98
    - (B) 1126.72
    - (C) 1131.11
    - (D) 1136.59
    - (E) 1141.34

Provide the answers in the spaces below:

6. _____

7. _____

8. _____

9. _____

10. _____

CONTINUE ON TO THE NEXT PAGE...

Problem 3

Platplan.Dwg

Directions for Platplan.Dwg

Start a new drawing called Platplan. Use the Units command to change the system of units from decimal to engineering; change the number of decimal places past the zero from 4 to 2. Change the system of angle measurement to Surveyors units. Keep the remaining Units command default values. No special limits need be set for this drawing. Do not add any dimensions to this drawing.

Use the detail measurements at the left to construct the house and all decks outlined by the crosshatching. The house is to be aligned parallel with line AB of the plat. Answer the questions on the next page regarding this drawing.

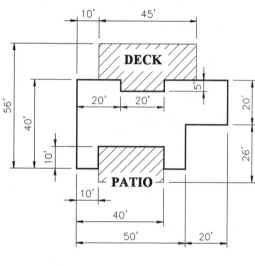

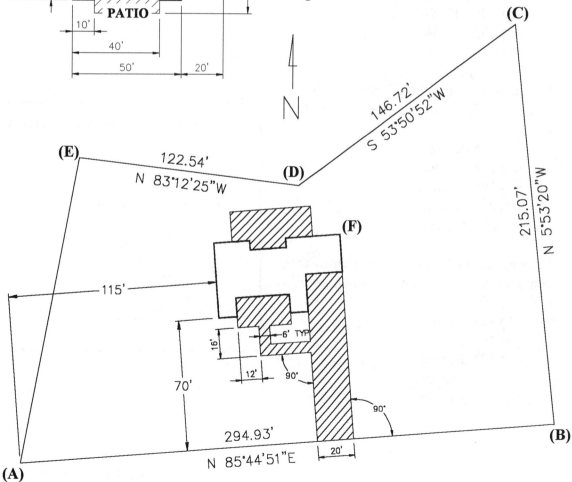

Questions for Platplan.Dwg

Refer to the drawing of the Platplan on the previous page to answer questions #11 through #15 below:

11. The correct direction of the line segment from the endpoint at "E" to the endpoint at "A" is
    (A) S 10d30'20"W
    (B) S 11d1'30"W
    (C) S 12d52'10"W
    (D) S 13d23'17"W
    (E) S 14d9'24"W

12. The total area of the plat with only the house removed is closest to
    (A) 42082.00 sq. ft.
    (B) 42086.00 sq. ft.
    (C) 42090.00 sq. ft
    (D) 42094.00 sq. ft.
    (E) 42098.00 sq. ft.

13. The total area of the deck, patio, sidewalks, and driveway is closest to
    (A) 3328.00 sq. ft.
    (B) 3332.00 sq. ft.
    (C) 3336.00 sq. ft.
    (D) 3340.00 sq. ft.
    (E) 3344.00 sq. ft.

14. The distance from the intersection of the corner of the house at "F" to the intersection of vertex "B" is closest to
    (A) 154'-6.00"
    (B) 154'-9.00"
    (C) 155'-0.00"
    (D) 155'-3.00"
    (E) 155'-6.00"

15. An error has been discovered in the original plat. Vertex "C" needs to be stretched straight down a distance of 35.00'. The total area of the plat with the house, deck, patio, sidewalks, and driveway removed is closest to
    (A) 36302.00 sq. ft.
    (B) 36306.00 sq. ft.
    (C) 36310.00 sq. ft.
    (D) 36314.00 sq. ft.
    (E) 36318.00 sq. ft.

Provide the answers in the spaces below:

11. _____

12. _____

13. _____

14. _____

15. _____

END OF SECTION I - DO NOT PROCEED FURTHER UNTIL TOLD TO DO SO

Notes

Level II Exam Post-Test #1 Section II General Knowledge Segment

Answer each of the 25 general knowledge questions. The questions may be answered in any order. It is considered good practice to answer the easier questions first. If a question seems difficult, do not waste time trying to answer it. Go on to the next question and come back to the difficult question or questions later. Be sure to provide the best possible answer for each question.

You should allow yourself a total of 30 minutes to answer all 25 general knowledge questions.

Place the best answer in the appropriate box for each of the following questions. Unless otherwise specified, all questions are "Single Answer Multiple Choice".

16. Command aliasing is found in the
   (A) ACAD.ALS file.
   (B) ACAD.PGP file.
   (C) ACAD.EXE file.
   (D) ALIAS.COM file.
   (E) ALIAS.BAT file.

Given the following layer names, answer the next question by supplying the appropriate range wildcard:

        FLOOR-1
        FLOOR-2
        FLOOR-3
        FLOOR-4
        FLOOR-5
        FLOOR-10
        FLOOR-11
        FLOOR-A
        FLOOR-B

17. In the group of layers illustrated above, to produce a list of all the layers that start with "Floor-" and are followed by two digits, use the following range wildcard:
   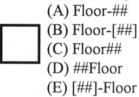
   (A) Floor-##
   (B) Floor-[##]
   (C) Floor##
   (D) ##Floor
   (E) [##]-Floor

Question 18 is considered a "Multiple-Answer Multiple-Choice question. Supply all possible answers that satisfy the particular question.

18. Valid modes of the ATTDEF command include
   (A) Invisible.
   (B) Constant.
   (C) Preset.
   (D) Verify.
   (E) Preview.

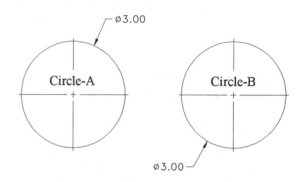

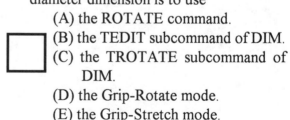

19. In the illustration above, the 3.00 diameter dimension in "Circle-A" needs to be relocated to a new position as in "Circle-B" without erasing the dimension and redimensioning the diameter. The best way to perform this repositioning of the diameter dimension is to use
   (A) the ROTATE command.
   (B) the TEDIT subcommand of DIM.
   (C) the TROTATE subcommand of DIM.
   (D) the Grip-Rotate mode.
   (E) the Grip-Stretch mode.

Question 20 is considered a "Short Answer or Fill In the Blank" question. Provide the correct answer in the space provided that satisfies the particular question.

20. File locking occurs when a user is prevented from editing a drawing because another user is currently working on the same drawing. A matching filename is created of the locked file with the extension

———————————————————

21. In the illustration above, the option of the PEDIT command that controls whether a linetype is generated from the start to the end of each polyline vertex or is generated across all vertices of the polyline is
    (A) Ltype.
    (B) Ltype Spline.
    (C) Ltype Gen.
    (D) Ltype Fit.
    (E) Splineframe.

22. The CMDDIA system variable controls
    (A) the display of the PLOT dialogue box.
    (B) the display of DDSELECT.
    (C) the display of DDINSERT.
    (D) All of the above.
    (E) Only A and C.

Question 23 is considered a "Multiple-Answer Multiple-Choice question. Supply all possible answers that satisfy the particular question.

23. Of the following names, those that are considered valid block names which may be inserted into a drawing include
    (A) CHAIR_1.
    (B) FURNITURE|CHAIR-1.
    (C) FURNITURE0CHAIR-1.
    (D) OFFICE0DOOR.
    (E) 3492A|DETAIL1.

24. Automatic paper space linetype scaling, the ability to scale a linetype in paper space based on the scale of the viewport in which it is scaled is controlled by the
    (A) PSPACE system variable.
    (B) PSLTSCALE system variable.
    (C) LTSCALE command.
    (D) LTYPE command.
    (E) SCALELTYPE system variable.

Question 25 is considered a "Multiple-Answer Multiple-Choice question. Supply all possible answers that satisfy the particular question.

25. Of the following operations, those that will help speed up the regeneration of a drawing include
    (A) turning QTEXT On.
    (B) freezing unused layers.
    (C) turning Off unused layers.
    (D) turning GRID Off.
    (E) turning GRID On.

CONTINUE ON TO THE NEXT PAGE...

26. Text can be edited using the DDEDIT dialogue box while in
    (A) Model Space.
    (B) Paper Space.
    (C) Text Space.
    (D) Both A and B.
    (E) None of the above.

27. Custom plot settings may be saved to
    (A) the plot configuration parameter file.
    (B) a .PCP file.
    (C) the ACAD.CFG file.
    (D) the ACAD.BAT file.
    (E) Only A and B.

Question 28 is considered a "Short Answer or Fill In the Blank" question. Provide the correct answer in the space provided that satisfies the particular question.

28. The system variable used to set an increment that automatically saves a drawing file to disk is

    _____

29. The system variable that controls whether or not the image of an entity is visible while it is being copied or moved is
    (A) VISMODE.
    (B) IMAGMODE.
    (C) REGENAUTO.
    (D) GHOSTMODE.
    (E) DRAGMODE.

30. The command used to read a drawing interchange file into an AutoCAD drawing is
    (A) DXFINTO
    (B) DXFIN
    (C) DXFREAD
    (D) DXFSEND
    (E) DXFTRANS

```
+-------------------------------------+
|      Entity Selection Settings      |
| Selection Modes                     |
| +---------------------------------+ |
| | [ ] Noun/Verb Selection         | |
| | [ ] Use Shift to Add            | |
| | [ ] Press and Drag              | |
| | [ ] Implied Windowing           | |
| |  +---------------------------+  | |
| |  | Default Selection Mode    |  | |
| |  +---------------------------+  | |
| +---------------------------------+ |
| Pickbox Size                        |
| +---------------------------------+ |
| |   Min         Max               | |
| |                              [] | |
| | [<] [ ]              [>]         | |
| +---------------------------------+ |
|      +-----------------------+      |
|      | Entity Sort Method... |      |
|      +-----------------------+      |
| +--------+  +--------+  +---------+ |
| |   OK   |  | Cancel |  | Help... | |
| +--------+  +--------+  +---------+ |
+-------------------------------------+
```

31. In the illustration of the DDSELECT dialogue box above, the option that allows the user to create a selection set first and then use an editing command to affect the selection set is
    (A) Noun/Verb Selection.
    (B) Use Shift to Add.
    (C) Press and Drag.
    (D) Implied Windowing.
    (E) Entity Sort Method . . .

32. Paper Space allows
    (A) plotting of multiple viewports.
    (B) display of drawings at different scales on the same drawing sheet.
    (C) insertion of such items as title blocks for annotating drawings.
    (D) All of the above.
    (E) Both A and C.

33. While in Model Space, the command used to set up multiple viewport windows is
    (A) VPORTS.
    (B) VIEWPORT.
    (C) VIEWPORTS.
    (D) Only A and C.
    (E) Only A and B.

34. To make a reference file a permanent part of the current drawing database, use the XREF command with the
    (A) "*" option.
    (B) Explode option .
    (C) Reload option.
    (D) Attach option.
    (E) Bind option.

35. Given the illustration of the region model above, the command used to create three holes in the region model is
    (A) TRIM.
    (B) REMOVE.
    (C) SUBTRACT.
    (D) ERASE.
    (E) VOID.

36. In the illustration above, when identifying a new user coordinate system, the icon updates to its new location after using the UCSICON command and the
    (A) New option.
    (B) Locate option.
    (C) On option.
    (D) Origin option.
    (E) Off option.

37. While in Paper Space, to make a layer visible in one viewport but invisible or frozen in all other viewports, use the
    (A) LAYER command.
    (B) VPLAYER command.
    (C) VIEWPORT command.
    (D) DDLMODES command.
    (E) Both B and D.

CONTINUE ON TO THE NEXT PAGE...

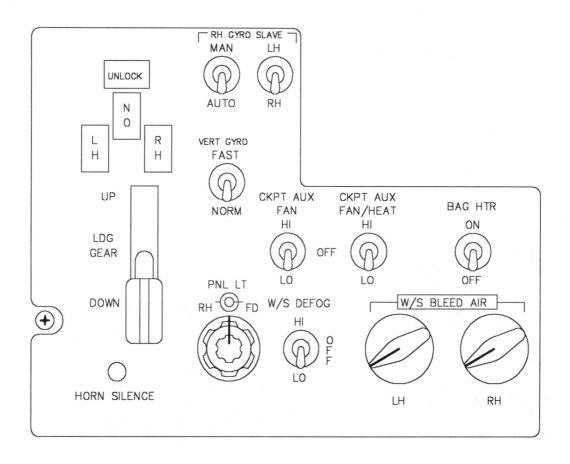

38. Given the illustration at the right of the control panel, to select all text that is 0.10 in height and to select all arcs, the list box of the Filter dialogue box also illustrated at the right should read

(A) Entity = Text
    Text Height = 0.10
    Entity = Arc

(B) **Begin OR
    **Begin AND
    Entity = Text
    Text Height = 0.10
    **End AND
    **Begin AND
    Entity = Arc
    **End AND
    **End OR

(C) Text
    Height = 0.10
    Arc

(D) **Begin
    Entity = Text
    Text Height = 0.10
    Entity = Arc
    **End

(E) **Begin AND
    **Begin OR
    Entity = Text
    Text Height = 0.10
    **End OR
    **Begin OR
    Entity = Arc
    **End OR
    **End AND

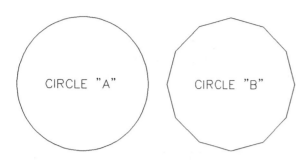

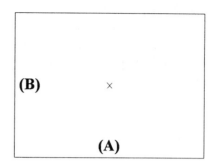

39. In the illustration above, the command that affects whether Circles and Arcs are displayed as a series of vectors as in Circle "B" which is used to produce fast zooms and lead to increased drawing productivity is

    (A) REGENAUTO.
    (B) VIEWRES.
    (C) CIRCLERES.
    (D) RESTORE.
    (E) ARCRES.

40. In the illustration above, the correct sequence to use for finding the centroid of the rectangle is

    (A) .XYZ filters using the following sequence:
        .X of the midpoint of "A"
        .Y of the midpoint of "B"
    (B) .XYZ filters using the following sequence:
        .X of the midpoint of "B"
        .Y of the midpoint of "A"
    (C) using the MEE expression of 'CAL
    (D) Only A and C
    (E) Only B and C

END OF SECTION II

UNIT

12

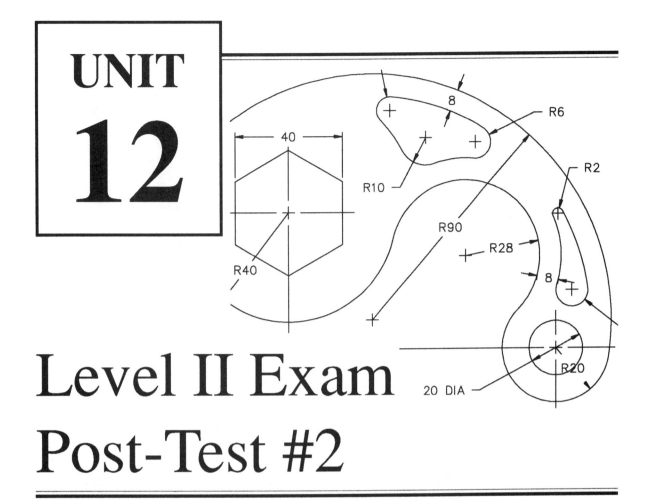

Level II Exam

Post-Test #2

Use this Level II Certification Exam Post-Test #2 for more practice in preparing for the full 3-hour AutoCAD Level II Certification Exam. This test is designed to be completed in 2 hours, which represents 2/3 actual exam. The AutoCAD Level II Certification Exam Post-Test #2 consists of the following two segments:

- The first segment concentrates on drawing skills. Three problems will be drawn and the questions answered in a total of 90 minutes.

- The second segment concentrates on general AutoCAD knowledge in the form of 25 general knowledge questions. These 25 questions are to be answered in 30 minutes.

Work through the Level II Post-Test #2 at a good pace paying strict attention to the amount of time spent on each drawing problem and general knowledge question. Answers for all Level II Post-Test #2 questions are located in Unit 13, page 228.

Notes

Level II Exam Post-Test #2 Section I Drawing Segment

Construct the three drawing problems assigned to this segment and answer the questions that follow each problem. The problems may be completed in any order. You should allow yourself a total of 90 minutes to complete all three problems.

When all problems have been completed and time still remains, use the extra time to carefully check your answers.

Problem 1

Ratchet.Dwg

Directions for Ratchet.Dwg

Start a new drawing called Ratchet. Keep the default units set to decimal units but change the number of decimal places past the zero from 4 to 2. Keep all remaining default Units values. No special limits need be set for this drawing. Begin this drawing with the center of the 1.00 radius arc at absolute coordinate (6.00,4.50). Do not add any dimensions to this drawing. Answer the questions on the next page regarding this drawing.

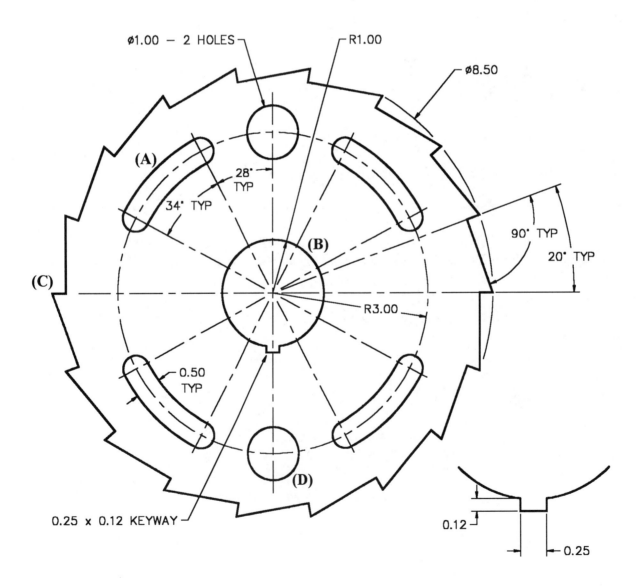

Questions for Ratchet.Dwg

Refer to the drawing of Ratchet on the previous page to answer questions #1 through #5 below:

1. The total length of arc "A" is
    (A) 1.81
    (B) 1.93
    (C) 2.08
    (D) 2.19
    (E) 2.31

2. The perimeter of the 1.00 radius arc "B" with the 0.25 x 0.12 keyway is
    (A) 6.52
    (B) 6.63
    (C) 6.77
    (D) 6.89
    (E) 7.02

3. The total surface area of the ratchet with all 4 slots, the two 1.00 diameter holes, and the 1.00 radius arc with the keyway removed is
    (A) 42.98
    (B) 43.04
    (C) 43.10
    (D) 43.16
    (E) 43.22

4. The absolute coordinate value of the endpoint at "C" is
    (A) 2.01,5.10
    (B) 2.01,5.63
    (C) 2.01,5.95
    (D) 2.20,5.95
    (E) 2.37,5.95

5. The angle formed in the X-Y plane from the endpoint of the line at "C" to the center of the 1.00 diameter hole "D" is
    (A) 300 degrees.
    (B) 303 degrees.
    (C) 306 degrees.
    (D) 309 degrees.
    (E) 312 degrees.

Provide the answers in the spaces below:

1. _____

2. _____

3. _____

4. _____

5. _____

CONTINUE ON TO THE NEXT PAGE...

Problem 2

Duplex.Dwg

Directions for Duplex.Dwg

Start a new drawing called Duplex. Begin the construction of the Duplex illustrated below by changing the system of units from decimal to architectural. Keep the remaining default unit settings. No special limits need be set for this drawing. Do not add any dimensions to this drawing. Begin drawing the lower left corner of the duplex identified by "X" at absolute coordinate 2'-4 1/2",5'-9 3/4". **All wall thicknesses measure 4".** Answer the questions on the next page regarding this drawing.

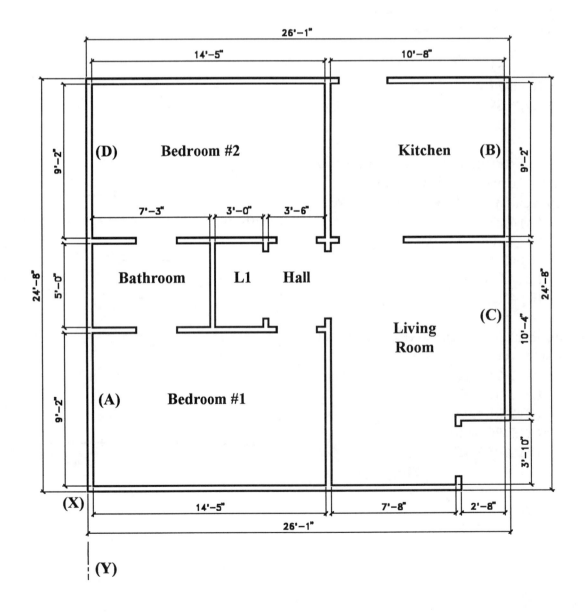

Questions for Duplex.Dwg

Refer to the drawing of the Duplex on the previous page to answer questions #6 through #10 below:

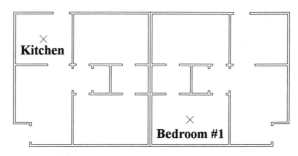

6. The distance from the midpoint of the inside wall "A" to the midpoint of the inside wall "B" is closest to
    (A) 28'-11"
    (B) 29'-2"
    (C) 29'-5"
    (D) 29'-8"
    (E) 29'-11"

7. The angle formed in the X-Y plane from the midpoint of the inside wall "C" to the midpoint of the inside wall "D" is closest to
    (A) 152 degrees.
    (B) 155 degrees.
    (C) 158 degrees.
    (D) 161 degrees.
    (E) 164 degrees.

8. The total combined area of Bedroom #1, Bedroom #2, the Bathroom, the Living Room, and the Laundry Room (L1) is closest to
    (A) 450 sq. ft.
    (B) 458 sq. ft.
    (C) 466 sq. ft.
    (D) 474 sq. ft.
    (E) 482 sq. ft.

9. The absolute coordinate value of the center of the Laundry Room (L1) is closest to
    (A) 11'-9 1/2",17'-4 3/4
    (B) 11'-9 1/2",17'-7 3/4
    (C) 11'-9 1/2",17'-10 3/4"
    (D) 11'-9 1/2",18'-1 3/4"
    (E) 12'-1 1/2",18'-4 3/4

10. Create a duplex apartment by using the MIRROR command to copy and flip all entities that make up the single apartment using the centerline "Y" in the illustration on the previous page as the mirror line. Your display should be similar to the illustration above. The total distance from the center of the kitchen located in the left duplex to the center of Bedroom #1 located in the right duplex is closest to
    (A) 31'-8"
    (B) 31'-11"
    (C) 32'-2"
    (D) 32'-5"
    (E) 32'-8"

Provide the answers in the spaces below:

6._____

7._____

8._____

9._____

10._____

CONTINUE ON TO THE NEXT PAGE...

Problem 3

Seal-1.Dwg

Directions for Seal-1.Dwg

Start a new drawing called Seal-1. Use the Units command to change the number of decimal places past the zero from 4 to 3. Keep the remaining Units command default values. No special limits need be set for this drawing. Do not add any dimensions to this drawing.

Begin constructing Seal-1 by locating the center of the 1.500 radius arc at absolute coordinate (5.000,5.875). Answer the questions on the next page regarding this drawing.

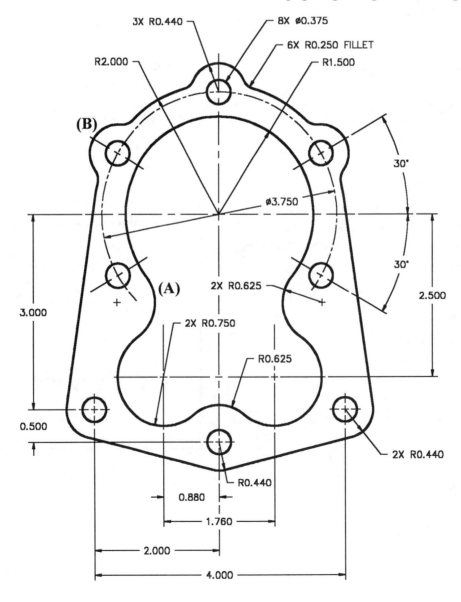

Questions for Seal-1.Dwg

Refer to the drawing of the Seal-1 on the previous page to answer questions #11 through #15 below:

11. The total length of arc "A" is
    (A) 1.042
    (B) 1.045
    (C) 1.048
    (D) 1.051
    (E) 1.054

12. The absolute coordinate value of the center of arc "B" is
    (A) 3.376,6.807
    (B) 3.376,6.810
    (C) 3.376,6.813
    (D) 3.379,6.813
    (E) 3.382,6.816

13. The area of the inside irregular shape of Seal-1 is
    (A) 12.023
    (B) 12.026
    (C) 12.029
    (D) 12.032
    (E) 12.035

14. The absolute coordinate value of the centroid of Seal-1 is closest to
    (A) 5.000,4.674
    (B) 5.000,4.677
    (C) 5.000,4.680
    (D) 5.000,4.683
    (E) 5.000,4.686

15. Change the diameter of the 5 circles located along the 3.750 diameter arc from the existing diameter of 0.375 to a new diameter of 0.560 units. The absolute coordinate location of the centroid is closest to
    (A) 5.000,4.553
    (B) 5.000,4.556
    (C) 5.000,4.559
    (D) 5.000,4.562
    (E) 5.000,4.565

Provide the answers in the spaces below:

11._____

12._____

13._____

14._____

15._____

END OF SECTION I - DO NOT PROCEED FURTHER UNTIL TOLD TO DO SO

Notes

Level II Exam Post-Test #2 Section II General Knowledge Segment

Answer each of the 25 general knowledge questions. The questions may be answered in any order. It is considered good practice to answer the easier questions first. If a question seems difficult, do not waste time trying to answer it. Go on to the next question and come back to the difficult question or questions later. Be sure to provide the best possible answer for each question.

You should allow yourself a total of 30 minutes to answer all 25 general knowledge questions.

Place the best answer in the appropriate box for each of the following questions. Unless otherwise specified, all questions are "Single Answer Multiple Choice".

Given the following layer names, answer the next question by supplying the appropriate range wildcard:

FLOOR-1
FLOOR-2
FLOOR-3
FLOOR-4
FLOOR-5
FLOOR-10
FLOOR-11
FLOOR-A
FLOOR-B

16. Using the group of layers illustrated above, to produce a list of layers "Floor-A" through "Floor-B", enter the following range wildcard:
    (A) Floor@
    (B) @Floor
    (C) @-Floor
    (D) Floor-@
    (E) @Floor@

17. The system variable that controls whether a dialogue box requesting file information is displayed or not displayed is
    (A) DDFILE.
    (B) FILEDIA.
    (C) FDIALOG.
    (D) DIALOG.
    (E) DDIA.

18. While in Paper Space, the command used to create user defined viewports is
    (A) VPORTS.
    (B) MVIEW.
    (C) VIEWPORTS.
    (D) Both A and C.
    (E) Both B and C.

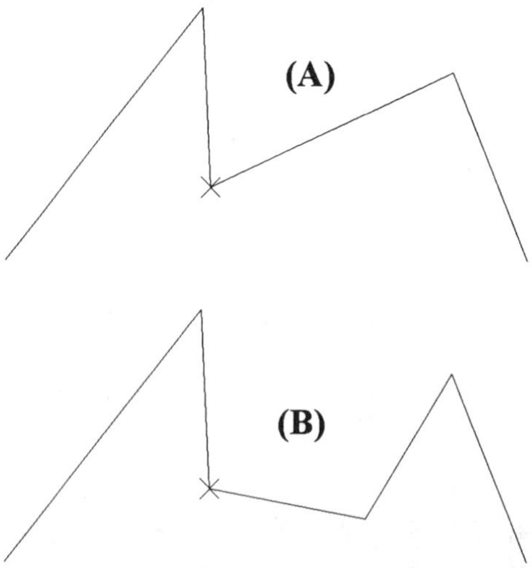

(A)

(B)

19. In the illustration above, the option of the PEDIT-Edit Vertex command used to change polyline entity "A" to polyline entity "B" is
    (A) Break.
    (B) Tangent.
    (C) Move.
    (D) Straighten.
    (E) Insert.

20. To perform fast zooms,
    (A) the REGEN command must be Off.
    (B) the ZOOM All option should be used.
    (C) the VIEWRES command must be On.
    (D) Only A and C.
    (E) None of the above.

Question 21 is considered a "Short Answer or Fill In the Blank" question. Provide the correct answer in the space provided that satisfies the particular question.

21. The command used to extract attributes into either the CDF, SDF, or DXF formats is

_____

22. Freeplotting allows for
    (A) plotting without using a network license.
    (B) plotting without waiting for a drawing regeneration.
    (C) plotting without creating any AutoCAD temporary files.
    (D) All of the above.
    (E) Only A and C.

23. To unlock a locked file
    (A) use the CONFIG command.
    (B) use the FILES command.
    (C) pick "Utilities..." from the "File" pulldown menu.
    (D) Both A and C.
    (E) Both B and C.

24. With TILEMODE set to 0, AutoCAD treats viewports like
    (A) named views created using the VIEW command.
    (B) those created using the VPORTS command.
    (C) other entities such as lines, circles, and arcs.
    (D) the Window option used in the ZOOM command.
    (E) the Dynamic option used in the ZOOM command.

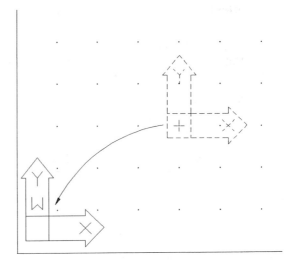

25. In the illustration above, the option of the UCS command that returns a user defined coordinate system back to the world coordinate system is
    (A) World.
    (B) Origin.
    (C) 0.50,0.50
    (D) 0,0
    (E) 3 point.

26. To change the dimension style of an existing associative dimension, use the
    (A) DIM-UPDATE command.
    (B) STRETCH command.
    (C) SCALE command.
    (D) All of the above.
    (E) None of the above.

27. The region modeling command used to preform such editing operations as scaling, moving, or repalacing individual primitives belonging to a region model is
    (A) EDITPRIM.
    (B) PRIMEDIT.
    (C) SOLCHP.
    (D) SOLEDIT.
    (E) SOLCHANGE.

CONTINUE ON TO THE NEXT PAGE...

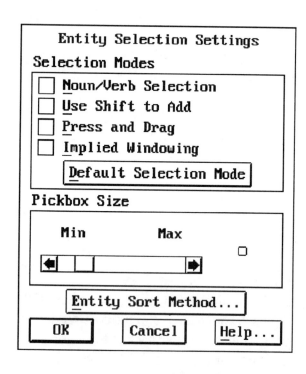

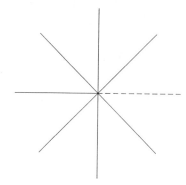

28. In the illustration of the DDSELECT dialogue box above, the option that allows the user to enter an editing command, pick a blank part of the screen, and automatically be either in window or crossing mode is

    (A) Noun/Verb Selection.
    (B) Use Shift to Add.
    (C) Press and Drag.
    (D) Implied Windowing.
    (E) Entity Sort Method . . .

29. External commands such as CATALOG and DIR or Command Aliases such as entering "L" for the LINE command are found in the

    (A) ACAD.EXE file.
    (B) ACAD.PGP file.
    (C) ACAD.COM file.
    (D) ACAD.HLP file.
    (E) ACAD.LIN file

30. Illustrated above is an object being constructed using the ARRAY command and selecting the highlighted entity. Without using the ARRAY command, select the **best** answer below as an alternate method of constructing the entities above:

    (A) Use the ROTATE command.
    (B) Using Grip-Rotate mode.
    (C) Using Grip-Rotate mode with the Multiple option.
    (D) Using Grip-Rotate mode with the Multiple option and the temporary snap mode through the "Shift" key.
    (E) None of the above.

31. To make a dependent object such as a block or layer belonging to an external reference file a permanent part of the current drawing file, use the

    (A) INSERT command.
    (B) XBIND command.
    (C) XREF command.
    (D) ATTACH command.
    (E) BIND command.

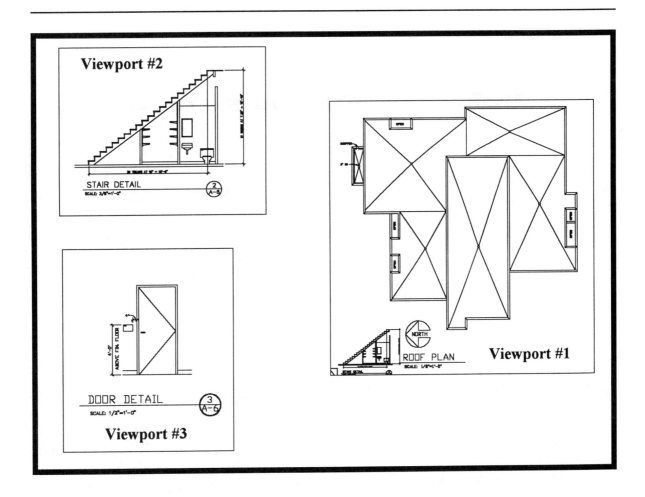

32. In the illustration above of the three viewports laid out in paper space, the command used to remove the unnecessary images of the stair detail and door detail from Viewport #1 is
    (A) ERASE.
    (B) PURGE.
    (C) MOVE.
    (D) SCALE.
    (E) VPLAYER.

CONTINUE ON TO THE NEXT PAGE...

Notes

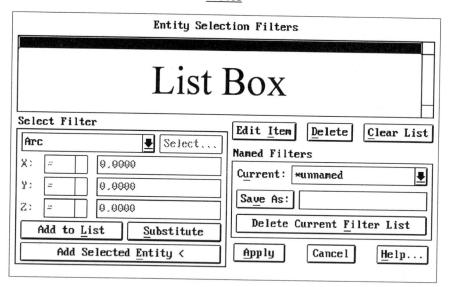

33. To select all circles in a drawing that have a radius of 1.00 units and to select all lines in the same drawing on layer "Object", the list box of the Filter dialogue box illustrated above should read

    (A) Entity = Circle
        Circle Radius = 1.00
        Entity = Line
        Layer = Object

    (B) **Begin AND
        **Begin OR
        Entity = Circle
        Circle Radius = 1.00
        **End OR
        **Begin OR
        Entity = Line
        Layer = Object
        **End OR
        **End AND

    (C) Circle
        Radius = 1.00
        Line
        Layer = Object

    (D) **Begin
        Entity = Circle
        Circle Radius = 1.00
        Circle Center X = 4.40 Y = 3.24
        Entity = Line
        Layer = Object
        **End

    (E) **Begin OR
        **Begin AND
        Entity = Circle
        Circle Radius = 1.00
        **End AND
        **Begin AND
        Entity = Line
        Layer = Object
        **End AND
        **End OR

Layer Name	State	Color	Linetype
0	On . . . . white		CONTINUOUS
1A6¦A-DIM-5	On 9		CONTINUOUS
1A6¦A-SYM-2	On . . . . blue		CONTINUOUS
1A6¦A-SYM-5	On 9		CONTINUOUS
1A6¦A-TX-1	On . . . . red		CONTINUOUS
1A6¦A-TX-5	On 9		CONTINUOUS
1A6¦C-ROOF-4	On F L C N green		CONTINUOUS
2A6¦A-DIM-1	On . . . . cyan		CONTINUOUS
2A6¦A-TX-1	On . . . . red		CONTINUOUS
2A6¦C-FLR-6	On 14		CONTINUOUS
2A6¦C-WAL-6	On 14		CONTINUOUS
2A6¦C-WAL-7	On 15		CONTINUOUS

34. In the illustration above of the layer dialogue box, for the highlighted layer "1A6|C-ROOF-4" the letters "C" and "N" located under the Layer State area of the dialogue box signifies

    (A) the layer is locked in the current viewport and in any new viewports.
    (B) the layer is turned on in the current viewport and in any new viewports.
    (C) the layer is turned off in the current viewport and in any new viewports.
    (D) the layer is thawed in the current viewport and in any new viewports.
    (E) the layer is frozen in the current viewport and in any new viewports.

CONTINUE ON TO THE NEXT PAGE...

```
1A6|A-DIM-5        On         9                  CONTINUOUS    Xdep: 1A6
1A6|A-SYM-2        On         5 (blue)           CONTINUOUS    Xdep: 1A6
1A6|A-SYM-5        On         9                  CONTINUOUS    Xdep: 1A6
1A6|A-TX-1         On         1 (red)            CONTINUOUS    Xdep: 1A6

1A6|A-TX-5         On         9                  CONTINUOUS    Xdep: 1A6
1A6|C-ROOF-4       On         3 (green)          CONTINUOUS    Xdep: 1A6
2A6|A-DIM-1        On         4 (cyan)           CONTINUOUS    Xdep: 2A6
2A6|A-TX-1         On         1 (red)            CONTINUOUS    Xdep: 2A6
2A6|C-FLR-6        On         14                 CONTINUOUS    Xdep: 2A6

2A6|C-WAL-6        On         14                 CONTINUOUS    Xdep: 2A6
2A6|C-WAL-7        On         15                 CONTINUOUS    Xdep: 2A6
2A6|M-MISC-7       On         15                 CONTINUOUS    Xdep: 2A6
3A6|A-DIM-5        On         9                  CONTINUOUS    Xdep: 3A6
3A6|A-SYM-1        On         1 (red)            CONTINUOUS    Xdep: 3A6

-- Press RETURN for more --
3A6|A-SYM-5        On         9                  CONTINUOUS    Xdep: 3A6
3A6|A-TX-1         On         1 (red)            CONTINUOUS    Xdep: 3A6
3A6|A-TX-5         On         9                  CONTINUOUS    Xdep: 3A6
```

35. Illustrated above is a listing of a series of layers associated with a particular drawing. The most efficient way to identify all layers assigned the color "Red" is to

    (A) visually see which layers have "Red" assigned to them and list these layers on a separate sheet of paper.

    (B) use the "Set Layer Filter" dialogue box illustrated at the right to change the "*" next to "Ltypes:" to "Hidden".

    (C) use the "Set Layer Filter" dialogue box illustrated at the right to change the "*" next to "Layer Names:" to "1A6|A-TX-1".

    (D) use the "Set Layer Filter" dialogue box illustrated at the right to change the "*" next to "Colors:" to "Red".

    (E) use the "Set Layer Filter" dialogue box illustrated at the right to change the "*" next to "Layer Names:" to "3A6|A-SYM-1".

Set Layer Filters

On/Off:	Both ▼
Freeze/Thaw:	Both ▼
Lock/Unlock:	Both ▼
Current Vport:	Both ▼
New Vports:	Both ▼
Layer Names:	*
Colors:	*
Ltypes:	*

Reset

| OK | Cancel | Help... |

```
┌──────────────────────────────────────────┐
│                 Colors                    │
│  Style: *UNNAMED                          │
│                                           │
│  Feature Scaling:      ┌──────────────┐   │
│                        │  1.00000     │   │
│  ☐ Use Paper Space Scaling                │
│                                           │
│  Dimension Line Color:  ┌──────┐  ▓▓▓▓    │
│                         │ red  │  ▓▓▓▓    │
│  Extension Line Color:  ┌──────┐  ▓▓▓▓    │
│                         │ red  │  ▓▓▓▓    │
│  Dimension Text Color:  ┌──────┐  ▓▓▓▓    │
│                         │yellow│  ▓▓▓▓    │
│   ┌──────┐   ┌──────────┐   ┌─────────┐   │
│   │  OK  │   │  Cancel  │   │ Help... │   │
│   └──────┘   └──────────┘   └─────────┘   │
└──────────────────────────────────────────┘
```

36. In the illustration above of the "DDIM-Colors" dialogue box, the purpose of assigning a different color to the dimension text would be to

    (A) make the drawing look more interesting with numerous colors.

    (B) override the purpose of controlling colors by layers.

    (C) assign a thicker pen to the color of the dimension text for it to plot out thicker than the extension or dimension lines.

    (D) explode the dimension and then change the color of the dimension text to an existing layer.

    (E) None of the above.

37. The file automatically updated whenever a digitizing tablet is configured, or whenever grips are turned on, or whenever Noun/Verb selection mode in the "Entity Selection Settings" dialogue box is checked is the

    (A) ACAD.PAT file.

    (B) ACAD.PGP file.

    (C) CONFIG.SYS file.

    (D) ACAD.EXE file.

    (E) ACAD.CFG file.

38. The option of the DRAGMODE command that allows the user to control whether a group of highlighted entities drags across the display screen for such commands as COPY and MOVE or does not drag is

    (A) Auto.

    (B) On.

    (C) Off

    (D) Normal.

    (E) Default.

39. You have just made a layer named "Entities" the new current layer. The color "Yellow" has already been assigned to this layer enabling you to draw "Yellow" colored entities. However, as you draw a few line segments, you notice the segments are actually being drawn in the "Magenta" color. This is due to

    (A) the wrong layer being made current.

    (B) the wrong linetype being assigned to the layer named "Entities".

    (C) the wrong color being assigned to the layer named "Entities".

    (D) the color "Magenta" being the new default color set by the COLOR command.

    (E) the color "Yellow" being the new default color set by the COLOR command.

CONTINUE ON TO THE NEXT PAGE...

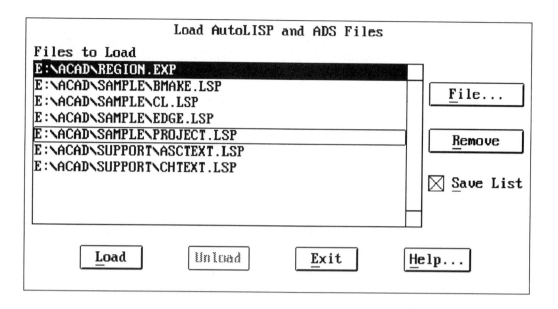

40. In the illustration of the APPLOAD dialogue box, the ability to remember previously used LISP or ADS routines is controlled by the

    (A) APPLOAD.DFS file.
    (B) ACAD.PGP file.
    (C) ACAD.EXE file.
    (D) APPLOAD.EXE file.
    (E) APPLOAD.XLG file.

END OF SECTION II

UNIT 13

Level II Exam Pre/Post-Test Answers

Use this unit to check your success after taking the Level II Pre-Test and Post-Tests. Once finished with all tests, time yourself again on the same tests and see if you either have completed the problems faster or have correctly answered difficult general knowledge questions.

Answers to the Level II Pre-Test
Drawing Segment
and
General Knowledge Segment

1.	D	11.	B, C, D
2.	B	12.	C
3.	E	13.	A, C, D
4.	D	14.	A
5.	A	15.	C
6.	E	16.	E
7.	D	17.	Attdef *
8.	A	18.	A
9.	D	19.	B
10.	C	20.	C
		21.	B, E
		22.	E

* Another acceptable answer for question #17 is DDATTDEF.

Answers to the Level II Post-Test #1
Drawing Segment
and
General Knowledge Segment

1. E
2. B
3. D
4. A
5. D
6. A
7. B
8. C
9. B
10. D
11. B
12. C
13. A
14. E
15. A

16. B
17. A
18. A, B, C, D
19. D
20. .Dwk*
21. C
22. A
23. A, C, D
24. B
25. A, B, D
26. D
27. E
28. Savetime
29. E
30. B

31. A
32. D
33. D
34. E
35. C
36. D
37. E
38. B
39. B
40. D

* Another acceptable answer for question #20 is Dwk

Answers to the Level II Post-Test #2
Drawing Segment
and
General Knowledge Segment

1. B
2. A
3. D
4. C
5. E
6. C
7. C
8. B
9. D
10. A
11. C
12. C
13. B
14. A
15. E

16. D
17. B
18. B
19. E
20. C
21. Attext*
22. D
23. E
24. C
25. A
26. D
27. C
28. D
29. B
30. D

31. B
32. E
33. E
34. E
35. D
36. C
37. E
38. B
39. D
40. A

* Another acceptable answer for question #21 is DDATTEXT

Appendix 1

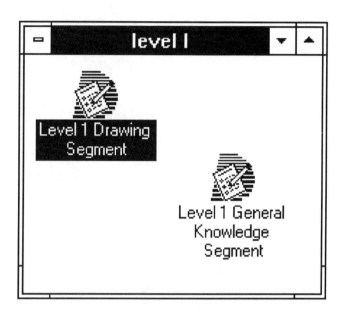

The Drake Training and Testing Electronic Delivery Mechanism

Notes

Components of a Typical Test Screen

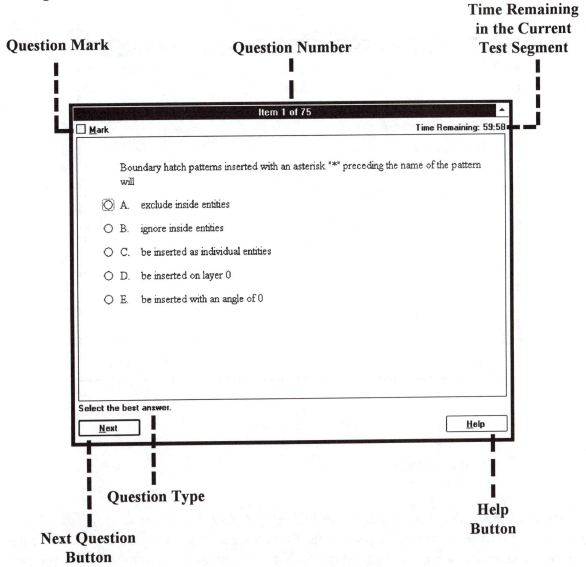

Question Mark

Question Number

Time Remaining in the Current Test Segment

Next Question Button

Question Type

Help Button

Illustrated above is a typical Drake Test Screen designed for all General Knowledge Segment questions of the AutoCAD Certification Exams. Components include the current question number, the ability to mark a question for later review, the time remaining for the current test segment, the question type, a button leading the individual to the next test question, and a help button providing information on items related to the operation of the test. In the center of the screen is the actual question followed by possible answers.

Single Answer Multiple Choice Test Screen

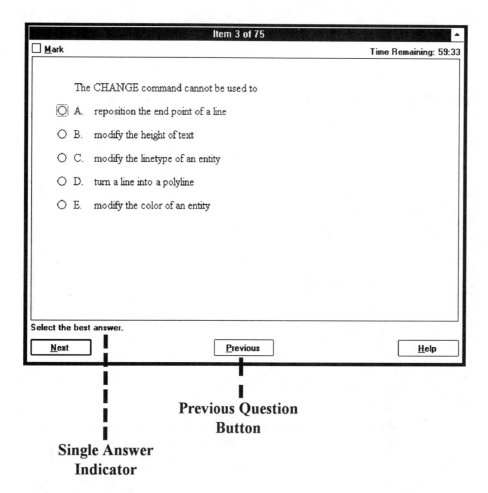

Use the illustration above as a guide for identifying a typical Single Answer Multiple Choice Question. Strict attention should be paid to the lower left corner of this window (just above the Next button) identifying this type of question. Only one answer may be accepted for this question type. Picking the appropriate circle places a dot in the center signifying the answer selected. If an individual realizes the wrong answer was selected, simply select another circle and the dot changes to the new circle. Once the first question has been answered, future questions display the Previous button enabling the individual to step back to previous questions.

Multiple Answer Multiple Choice Test Screen

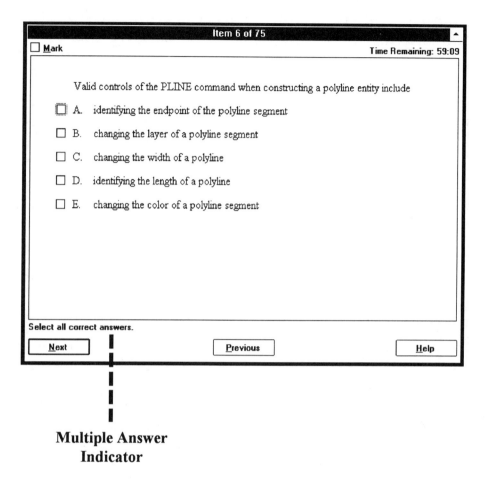

**Multiple Answer
Indicator**

The Multiple Answer Multiple Choice question indicator is illustrated above. The individual is responsible for supplying all possible answers to the current question. As each box is picked, a check appears; if the wrong box was mistakenly checked, pick the box a second time and the check disappears.

Short Answer (Fill in the Blank) Test Screen

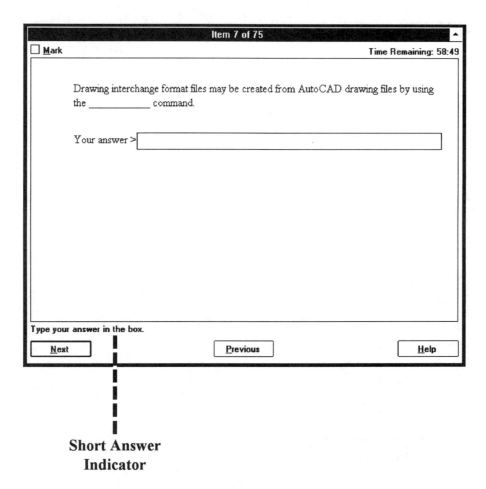

**Short Answer
Indicator**

Illustrated above is an example of a Short Answer or Fill in the Blank question. The correct answer must be entered in the edit box from the keyboard; in most cases spelling is very important.

Item Review Screen

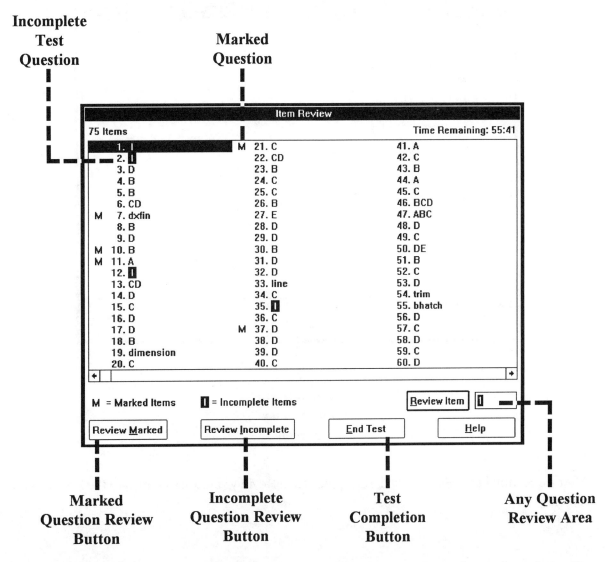

The Item Review Window is very important in reviewing either certain or all questions depending on the time remaining. If the Mark box was checked while in a question, these questions are identified in the Item Review Screen by the letter "M". An individual may either select these questions individually or may select the "Review Marked" button in the lower left corner of the window. All marked questions will be redisplayed enabling the individual to keep the current answer or make changes. If a question was mistakenly skipped, the letter "I" will appear next to the question signifying an incomplete question. Selecting the "Review Incomplete" button will step through all questions that have not been answered. Any test question may be reviewed by entering the question number in the edit box and selecting the "Review Item" button. When all questions have been reviewed, the individual may select the "End Test" button which will exit the test segment and officially record all answers.

Typical Question Review Screen

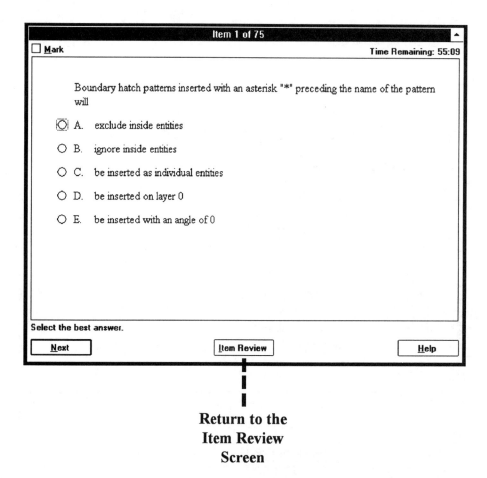

**Return to the
Item Review
Screen**

After a question has been reviewed, select the "Item Review" button to return to the Item Review screen.

Typical Drawing Segment Test Screen

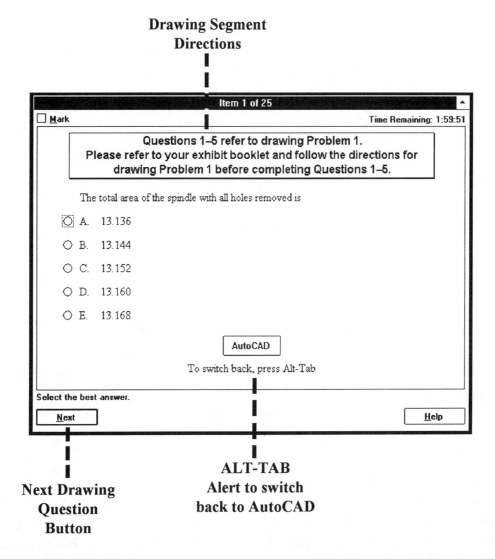

**Drawing Segment
Directions**

Item 1 of 25

☐ **M**ark   Time Remaining: 1:59:51

**Questions 1–5 refer to drawing Problem 1.
Please refer to your exhibit booklet and follow the directions for
drawing Problem 1 before completing Questions 1–5.**

The total area of the spindle with all holes removed is

○ A.   13.136

○ B.   13.144

○ C.   13.152

○ D.   13.160

○ E.   13.168

AutoCAD

To switch back, press Alt-Tab

Select the best answer.

Next   **H**elp

**Next Drawing
Question
Button**

**ALT-TAB
Alert to switch
back to AutoCAD**

All Drawing Segment screens are very similar to the previously discussed General Knowledge Segment screens with a few exceptions. Currently, all Drawing Segment questions are designed to be Single Answer Multiple Choice; in other words, only one answer will be accepted. The large title strip at the top of this screen alerts the individual to the drawing problem and the questions associated with it. All drawings will be performed using the latest version of AutoCAD for Windows. When confronted with a question, the individual must be comfortable with re-entering the AutoCAD environment to answer the question; This is accomplished by simultaneously pressing the ALT and TAB buttons. This automatically switches the user back to the AutoCAD drawing to perform an Inquiry command operation depending on the question. Simultaneously pressing ALT and TAB inside of the AutoCAD drawing editor switches back to the question for it to be answered.